This report contains the collective views of an international group of experts and does not necessarily represent the decisions or the stated policy of either the World Health Organization or the United Nations Environment Programme.

AF822993

Environmental Health Criteria 15

TIN AND ORGANOTIN COMPOUNDS: A PRELIMINARY REVIEW

Published under the joint sponsorship of the United Nations Environment Programme and the World Health Organization

World Health Organization
Geneva, 1980

ISBN 92 4 154075 3

© World Health Organization 1980

Publications of the World Health Organization enjoy copyright protection in accordance with the provisions of Protocol 2 of the Universal Copyright Convention. For rights of reproduction or translation of WHO publications, in part or *in toto*, application should be made to the Office of Publications, World Health Organization, Geneva, Switzerland. The World Health Organization welcomes such applications.

The designations employed and the presentation of the material in this publication do not imply the expression of any opinion whatsoever on the part of the Secretariat of the World Health Organization concerning the legal status of any country, territory, city or area or of its authorities, or concerning the delimitation of its frontiers or boundaries.

The mention of specific companies or of certain manufacturers' products does not imply that they are endorsed or recommended by the World Health Organization in preference to others of a similar nature that are not mentioned. Errors and omissions excepted, the names of proprietary products are distinguished by initial capital letters.

PRINTED IN FINLAND

79/4569 — VAMMALAN KIRJAPAINO OY, VAMMALA — 7000

CONTENTS

TIN AND ORGANOTIN COMPOUNDS: A PRELIMINARY REVIEW

NOTE TO READERS OF THE CRITERIA DOCUMENTS

While every effort has been made to present information in the criteria documents as accurately as possible without unduly delaying their publication, mistakes might have occurred and are likely to occur in the future. In the interest of all users of the environmental health criteria documents, readers are kindly requested to communicate any errors found to the Division of Environmental Health, World Health Organization, Geneva, Switzerland, in order that they may be included in corrigenda which will appear in subsequent volumes.

In addition, experts in any particular field dealt with in the criteria documents are kindly requested to make available to the WHO Secretariat any important published information that may have inadvertently been omitted and which may change the evaluation of health risks from exposure to the environmental agent under examination, so that the information may be considered in the event of updating and re-evaluation of the conclusions contained in the criteria documents.

WHO TASK GROUP ON ENVIRONMENTAL HEALTH ASPECTS OF TIN AND ORGANOTIN COMPOUNDS

Members

Professor R. Lauwerys, Unité de Toxicologie industrielle et médicale, Université catholique de Louvain, Bruxelles, Belgium *(Rapporteur)*

Dr J. G. Noltes, Department of Organic and Organo-Element Chemistry, Institute for Organic Chemistry, Organization for Applied Scientific Research (TNO), Utrecht, Netherlands

Professor M. Piscator, Department of Environmental Hygiene, The Karolinska Institute, Stockholm, Sweden

Professor A. B. Roščin, Department of Occupational Hygiene, Central Institute for Advanced Medical Training, Moscow, USSR

Dr M. Sharratt, Division of Environmental Health & Chemical Hazards, Department of Health & Social Security, London, England *(Chairman)*

Professor M. Timar, State Institute of Occupational Health, Budapest, Hungary *(Vice-Chairman)*

Dr L. Čemišanska, Department of Occupational Toxicology, Centre of Hygiene, Sofia, Bulgaria

Secretariat

Dr L. Fishbein, National Centre for Toxicological Research, US Food & Drug Administration, Jefferson, AR, USA *(Temporary Adviser)*

Dr Y. Hasegawa, Control of Environmental Pollution and Hazards, Division of Environmental Health, WHO, Geneva, Switzerland

Dr J. E. Korneev, Control of Environmental Pollution and Hazards, Division of Environmental Health, WHO, Geneva, Switzerland

Dr A. Stiles, Division of Vector Biology and Control, WHO, Geneva, Switzerland

Dr V. B. Vouk, Control of Environmental Pollution and Hazards, Division of Environmental Health, WHO, Geneva, Switzerland *(Secretary)*

The following list of organotin compounds includes the trivial name of the compound used throughout the document, the Chemical Abstracts Service (CAS) name and number, the molecular formula, and alternative names.

ORGANOTIN COMPOUNDS

Name used in text	CAS Index name	CAS number	Molecular formula	Synonyms
Monosubstituted compounds				
ethyltin trichloride	stannane, trichloroethyl-(9CI) (8CI)	1066-57-5	$C_2H_5Cl_3Sn$	trichloroethylstannane; trichloroethyltin; ethyltrichlorostannane; ethyltrichlorotin
ethyltin triiodide	stannane, ethyltriiodo- (9CI) (8CI)	3646-94-46	$C_2H_5I_3Sn$	triiodoethyltin
butyltin trichloride	stannane, butyltrichloro- (9CI) (8CI)	1118-46-3	$C_4H_9Cl_3Sn$	monobutyltin trichloride; trichlorobutyltin; butyltrichlorostannane; butyltrichlorotin; trichlorobutyltin; trichlorobutylstannane; stannane, trichlorobutyl-
butylstannoic acid	stannane, butylhydroxyoxo- (9CI) (8CI)	2273-43-0	$C_4H_{10}O_2Sn$	1-butanestannoic acid; butyltin hydroxide oxide; butylstannoic acid (VAN)
butylthiostannoic acid	stannane, butylmercaptooxo- (8CI)	26410-42-4	$C_4H_{10}O\ S\ Sn$	
butyltin-S,S′,S″-tris (isooctylmercaptoacetate)	acetic acid, 2,2′,2″-((butylstannylidyne)) tris (thio) tris-, triisooctyl ester (9CI); acetic acid, ((butylstannylidyne) trithio) tri,-triisooctylester (8CI)	25852-70-4	$C_{34}H_{66}O_6S_3Sn$	butyltin tris(isooctyl thioglycolate); stannane, butyltris((carboxymethyl) thio)-, triisooctyl ester; butyltin tris(isooctyl thioglycollate); monobutyltin tris(isooctylthioglycolate); monobutyltin tris(isooctyltin thioglycolate); butyltin tris(isooctyl mercaptoacetate); triisooctyl(butylstannylidyne) trithio) triacetate; butylstannane tris(isooctyl mercaptoacetate)
butyltin-S,S′,S″-tris (2-ethylhexylmercaptoacetate)	8 oxa-3,5-dithia-4-stannatetradecanoic acid, 4 butyl-10-ethyl-4-((2-((2-ethylhexyl)oxyl)-2-oxoethyl) thio)-7-oxo-, 2-ethylhexyl ester (9CI); acetic acid, ((butylstannylidyne) trithio)tri-,tris(2-ethylhexyl) ester (8CI)	26864-37-9	$C_{34}H_{66}O_6S_3Sn$	monobutyltin tris(2-ethylhexyl thioacetate), monobutyltin tris(2-ethylhexyl thioglycolate)
butyltin sulfide	distannathiane, dibutyldithioxo- (9CI); distannthiane, 1,3-dibutyl-1,3-dithioxo- (8CI)	15666-29-2	$C_8H_{18}S_3Sn_2$	butyl thiostannoic anhydride; monobutyltin sulfide
octyltin trichloride	stannanne, trichlorooctyl- (9CI) (8CI)	3091-25-6	$C_8H_{17}Cl_3Sn$	trichlorooctylstannane; octyltrichlorostannane; *n*-octyltin trichloride; *n*-octyltrichlorostannane; mono-*n*-octyltin trichloride; trichloro-*n*-octylstannane
octyltin tris(2-ethyl hexylmercaptoacetate)	8-oxa-3,5-dithia-4-stannatetradecanoic acid, 10-ethyl-4-((2-((2-ethylhexyl)oxy)-2-oxoethyl) thio)-4-octyl-7-oxo-, 2-ethylhexyl ester (9CI); acetic acid, ((octylstannylidyne) trithio)tri-, tris(2-ethylhexyl) ester (8CI)	27107-89-7	$C_{38}H_{74}O_6S_3Sn$	

cyclohexylstannoic acid	stannane, cyclohexylhydroxyoxo- (9Cl) (8Cl)	22771-18-2	$C_6H_{12}O_2Sn$	cyclohexanestannoic acid
Disubstituted compounds				
dimethyltin dichloride	stannane, dichlorodimethyl- (9Cl) (8Cl)	753-73-1	$C_2H_6Cl_2Sn$	dichlorodimethylstannane; dichlorodimethyltin; dimethyldichlorostannane
dimethyltin S,S′-bis (isooctyl mercaptoacetate)	acetic acid, 2,2′-((dimethylstannylene) bis (thio))bis-, diisooctyl ester (9Cl)	26636-01-1	$C_{22}H_{44}O_4S_2Sn$	TM 181; diisooctyl ((dimethylstannylene) dithio)diacetate; dimethyltin bis(isooctyl thioglycolate); dimethyltin-*S,S′*-bis(isooctylthyoglycolate); bis((((isooctyloxy)-carbonyl)methyl)thio) dimethyltin; T 40 (ester); TM 181S; Advastab TM 181S; Advastab TM 181 S
diethyltin dichloride	stannane, dichlorodiethyl-	866-55-7	$C_4H_{10}Cl_2Sn$	tin, dichlorodiethyl-; dichlorodiethyltin; diethyldichlorotin; dichlorodiethylstannane; diethyldichlorostannane
diethyltin diiodide	stannanne, diethyldiiodo-	2767-55-7	$C_4H_{10}I_2Sn$	tin, diethyldiiodo-; diethyldiiodostannane
diethyltin dioctanoate	stannane, diethylbis((1-oxo octyl)oxy) (9Cl); stannane, diethylbis (octanoyloxy) (8Cl)	2641-56-7	$C_{20}H_{40}O_4Sn$	diethyltin dicaprylate
dipropyltin dichloride	stannane, dichlorodipropyl- (9Cl) (8Cl)	867-36-7	$C_6H_{14}Cl_2Sn$	dipropyltin chloride; dichlorodipropylstannane; dichlorodipropyltin; di-*n*-propyltin dichloride
diisopropryltin dichloride	stannane, dichlorobis (1-methylethyl)- (9Cl)	38802-82-3	$C_6H_{14}Cl_2Sn$	
dibutyltin dichloride	stannane, dibutyldichloro- (9Cl) (8Cl)	683-18-1	$C_8H_{18}Cl_2Sn$	dichlorodibutyltin; dibutyltin chloride; dichlorodibutylstannane; dibutyldichlorotin; di-*n*-butyltin dichloride; dibutyldichlorostannane
dibutyltin oxide	stannane, dibutyloxo- (9Cl) (8Cl)	818-08-6	$C_8H_{18}OSn$	di-*n*-butyltin oxide; tin, dibutyloxo-; dibutylstannane oxide; dibutyloxotin; dibutyloxostannane
dibutyltin diacetate	stannane, bis(acetyloxy)dibutyl- (9Cl)	1067-33-0	$C_{12}H_{24}O_4Sn$	diacetoxydibutyltin; T 1[catalyst]; T 1 (VAN); Ba 2726
dibutyltin dilaurate	stannane, dibutylbis ((1-oxododecyl)oxy)-	77-58-7	$C_{32}H_{64}O_4Sn$	Butynorate; DBTL; dibutylbis(lauroyloxy)-tin; Stabilizer D-22; tin dibutyl dilaurate; Stanclere DBTL; Davainex; TVS Tin Lau; dibutyltin didodecanoate; dibutylbis-(laurato)tin; Mark 1038; Tinostat; dibutyltin *n*-dodecanoate; Stavinor 1200 SN; T 12 (catalyst); dibutylstannylene dilaurate; T 12 (VAN)
dibutyltin maleate	1,3,2-dioxastannepin-4,7-dione, 2,2′-dibutyl-(9Cl) (8Cl)	78-04-6	$C_{12}H_{20}O_4Sn$	dibutyl(maleoyldioxy)tin; Advastab DBTM; Advastab T 290; Stavinor 1300 SN; dibutylstannylene maleate; Advastab T 340; Nuodex V 1525; Irgastab T 290
dibutyltin sulfide	stannane, dibutylthioxo- (9Cl) (8Cl)	4253-22-9	$C_8H_{18}SSn$	tin dibutyl mercaptide
dibutyltin di (2-ethylhexoate)	stannane, dibutylbis ((2-ethyl-1-oxohexyl)oxy)- (9Cl)	2781-10-4	$C_{24}H_{48}O_4Sn$	dibutyltin bis(2-ethylhexanoate); dibutyltin bis(alpha-ethylhexanoate)

Name used in text	CAS name	CAS number	Molecular formula	Synonyms
dibutyltin dioctanoate	stannane, dibutyl-bis((1-oxooctyl)oxy)- (9CI); stannane, dibutyl bis(octanoyloxy)- (8CI)	4731-77-5	$C_{24}H_{48}O_4Sn$	dibutyltin dioctoate; dibutyltin dicaprylate; dibutyltin octanoate
dibutyltin di (butyl maleate)	5,7,12-trioxa-6-stannahexa deca-2,9-dienoic acid,6,6-dibutyl-4,8,11-trioxo, butyl ester, (Z,Z)- (9CI) stannane, dibutylbis(3-carboxyacryloyl)oxy-, dibutyl ester,(Z,Z)-(8CI)	15546-16-4	$C_{24}H_{40}O_8Sn$	dibutyltin bis(monobutyl maleate); maleic acid dibutyltin salt (2:1) diisobutyl ester; B5 [stabilizer]; dibutyltin bis(butyl maleate)
dibutyltin di (nonylmaleate)	stannane, dibutylbis((3-carboxyacryloyl)oxy) -dinonyl ester, (Z,Z)- (8CI)	10584-97-1	$C_{34}H_{60}O_8Sn$	
dibutyltin *β*-mercapto propanoate	6H-1,3,2-oxathiastannin-6-one, 2,2-dibutyldihydro-	78-06-8	$C_{11}H_{22}O_2S\ Sn$	Thermolite 35; Advastab T-306; Mark 488; dibutyltin *O,S'*-mercaptopropionate, dibutyltin *S,O*-mercaptopropionate; dibutyltin *β*-mercaptopropionate
dibutyltin bis(lauryl mercaptide)	stannane, dibutylbis (dodecylthio)-(9CI) (8CI)	1185-81-5	$C_{32}H_{68}S_2Sn$	dibutylbis(dodecylthio)tin; Mellite 39; dibutyltin bis(dodecylmercaptide); bis-(dodecylthio)dibutyltin; dibutylbis(dodecyl-thio)dibutyltin; dibutylbis(dodecylthio)-stannane; Advastab TM 98; Mellite 139; Thermolite 20; dibutyltin *S,S'*-bis(dode-cylmercaptide)
dibutyltin "laurate-maleate"	2-butenoic acid, 4,4'-[(dibutyl-stannylene)bis(oxy)] bis [4-oxo-, (Z,Z)- mixed with dibutyl bis [(1-oxododecyl)oxy] stannane (1 : 1) (9CI)	73246-84-1	$C_{32}H_{64}O_4Sn$ $C_{16}H_{24}O_8Sn$	solution of dibutyltin dilaurate and dibutyltin maleate; Thermolite 17
dibutyltin *S,S'*-bis (isooctylthioglycolate)	acetic acid, 2,2' (dibutyl stannylene)bis(thio)bis-, diisooctyl ester (9CI); acetic acid, ((dibutylstannylene) dithio)di, diisooctyl ester (8CI)	25168-24-5	$C_{28}H_{56}O_4S_2Sn$	dibutyltin bis(isooctyl mercaptoacetate); dibutyltin *S,S'*-bis(isooctyl mercapto-acetate); diisooctyl((dibutylstannylene)-dithio)diacetate; Thermolite 31; bis(iso-octyloxycarbonylmethylthiolato) dibutyltin; dibutylbis((((isooctyloxy)carbonyl)methyl)-thio)tin; BTS 70; T 101 (accelerator); Irgastab 17M; T 101
dibutyltin *S,S'*-bis(2-ethylhexylmercaptoacetate)	8-oxa- 3,5-dithia-4-stannatetra decanoic acid 4,4-dibutyl-10-ethyl-7-oxo-2-ethylhexyl ester (9CI); acetic acid, ((dibutylstannylene) dithio)di-, bis(2-ethylhexyl) ester (9CI)	10584-98-2	$C_{28}H_{56}O_4S_2Sn$	dibutyltin bis(2-ethylhexyl thioglycolate); dibutyltin *S,S'*-bis(2-ethylhexyl thio-glycolate)
dipentyltin dichloride	stannane, dichlorodipentyl- (9CI)	1118-42-9	$C_{10}H_{22}Cl_2Sn$	dichlorodipentyltin
dioctyltin dichloride	stannane, dichlorodioctyl- (9CI) (8CI)	3542-36-7	$C_{16}H_{34}Cl_2Sn$	dichlorodioctyltin; dichlorodioctyl stannane
dioctyltin oxide	stannane, dioctyloxo- (9CI) (8CI)	870-08-6	$C_{16}H_{34}O\ Sn$	tin, dioctyloxo-; di-*n*-octyltin oxide; dioctyloxostannane
dioctyltin acetate	stannane, bis(acetyloxy)dioctyl- (9CI)	17586-94-6	$C_{20}H_{40}O_4Sn$	dioctyldiacetoxytin

dioctyltin dilaurate	stannane, dioctylbis((1-oxo dodecyl)oxy)-(9Cl); stannane, bis(lauroyloxy) dioctyl-(8Cl)	3648-18-8	$C_{40}H_{80}O_4Sn$	lauric acid, dioctyltin deriv; dioctyl-dilauroyloxytin; dioctyltin didodecanoate
dioctyltin maleate	1,3,2-dioxastannepin-4,7-dione, 2,2-dioctyl- (9Cl);	16091-18-2	$C_{20}H_{36}O_4Sn$	di-*n*-octyltin maleate; Thermolite 813; Estabex U 18; dioctylstannylene maleate; Mellite 825
dioctyltin dibutylmaleate	5,7,12-trioxa-6-stannahexadeca-2,9-dienoic acid, 6,6-dioctyl-4,8,11-trioxo-, butyl ester, (Z,Z)-(9Cl); stannane, bis(3-carboxyacryloyl) oxy)dioctyl-, dibutyl ester (Z,Z)-(8Cl)	29575-02-8	$C_{32}H_{56}O_8Sn$	dioctyltinbis(butylmaleate)
dioctyltin-*S*,*S′*-(ethylene glycol-bis-mercaptoacetate)	1,3-dioxa-6,9-dithia-2-stanna cycloundecane-4,11-dione,2,2-dioctyl-(9Cl)	56875-68-4	$C_{22}H_{42}O_4S_2Sn$	ethylenebisthioglycolate dioctyltin
dioctyltin-*S*,*S′*-bis(isooctyl-mercaptoacetate)	acetic acid, 2,2′-((dioctylstannylene)bis(thio)bis-,diisooctyl ester (9Cl); acetic acid, ((dioctylstannylene) dithio)di-, diisooctyl ester (8Cl)	26401-97-8	$C_{36}H_{72}O_4S_2Sn$	diisooctyl((dioctylstannylene)dithio)-diacetate; dioctyltin bis(isooctylmercapto-acetate); Thermalite 831; dioctyltin di(iso-octylthyoglycolate); di-*n*-octyltin bis(iso-octylmercaptoacetate), bis ((((isooctyloxy)-carbonyl) methyl)thio)dioctyltin; Mellite 831C; Irsastab 17 MOK; dioctyltin bis(iso-octylthyoglycolate); dioctyltin bis(diiso-octylthioglycolate)
dioctyltin mercaptoacetate	1,3,2-oxathiastannolan-5-one, 2,2-dioctyl-(9Cl) (8Cl)	15535-79-2	$C_{18}H_{36}O_2S\ Sn$	dioctyltin *S*,*O*-mercaptoacetate
dioctyltin *β*-mercapto propanoate	6H-1,3,2-oxathiastannin-6-one, dihydro-2,2-dioctyl- (9Cl) (8Cl)	3033-29-2	$C_{19}H_{38}O_2S\ Sn$	dioctyltin *S*,*O*-3-mercaptopropionate; dioctyltin mercaptopropionate
dioctyltin-*S*,*S′*-bis (butyl mercaptoacetate)	8-oxa-3,5-dithia-4-stanna dodecanoic acid, 4,4-dioctyl-7-oxo-, butyl ester (9Cl) acetic acid, (((dioctylstannylene) dithio)di-, butyl ester (8Cl)	27107-88-6	$C_{28}H_{56}O_4S_2Sn$	
dioctyltin-*S*,*S′*-bis(2-ethylhexylmercaptoacetate)	8-oxa-3,5-dithia-4-stannatetra decanoic acid, 10-ethyl-4,4-dioctyl-7-oxo-, 2-ethylhexyl ester (9Cl); acetic acid, ((dioctylstannylene) dithio)di-, bis(2-ethylhexyl) ester (8Cl)	15571-58-1	$C_{36}H_{72}O_4S_2Sn$	bis(2-ethylhexyl) (dioctylstannylene) dithio)diacetate; dioctyltin bis(2-ethyl-hexyl thioglycolate); bis(2-ethylhexylthyo-glycolato)dioctyltin; Advastab 17 MOK
dioctyltin-*S*,*S′*-bis (laurylmercaptoacetate)	8-oxa-3,5-dithia-4-stannaeicosanoic acid, 4,4-dioctyl-7-oxo, dodecyl ester (9Cl)	73246-85-2	$C_{44}H_{88}O_4S_2Sn$	dioctyltin bis(laurylthioglycolate)
dioctyltin bis(2-ethylhexylmaleate)	5,7,12-trioxa-6-stannaoctadeca-2,9-dienoic acid, 14-ethyl-6,6-dioctyl-4,8,11-trioxo, 2-ethylhexyl ester, (Z,Z) (9Cl)	10039-33-5	$C_{40}H_{72}O_8Sn$	di-*n*-octyltin bis(2-ethylhexylmaleate)

Name used in text	CAS name	CAS number	Molecular formula	Synonyms
dioctyltin bis(dodecyl mercaptide)	stannane, bis(dodecylthio) dioctyl-(9CI) (8CI)	22205-30-7	$C_{40}H_{84}S_2Sn$	bis(dodecylthio)dioctyltin; dioctyltin bis(lauryl mercaptide); dioctyltin, dilauryl mercaptan salt
dioctyltin-*S*,*S*'-(1,4-butanediol-bis-mercapto acetate)	1,9-dioxa-4,6-dithia-5-stannacyclotridecane-2,8-dione, 5,5-dioctyl- (9CI)	69226-46-6	$C_{24}H_{46}O_4S_2Sn$	
dioctyltin di(1,2-propyleneglycolmaleate)	1,3,8,11-tetraoxa-2-stannacyclopentadeca-5,13-diene-4,7,12,15-tetrone, 9-methyl-2,2-dioctyl-, (Z,Z)-(9CI)	69226-45-5	$C_{27}H_{44}O_8Sn$	
dioctyltin bis(isobutyl maleate)	stannane, bis[(3-carboxy-acryloyl)oxy]dioctyl-, diisobutyl ester, (Z,Z)- (9CI)	15571-59-2	$C_{32}H_{56}O_8Sn$	dioctyltin bis(isobutylmaleato)tin; 5,7,12-triox-6-stannapenta-2,9-dienoic acid, 6,6-dioctyl-13-methyl-4,8,11-trioxo, 1-methyl-propyl ester
diphenyltin dichloride	stannane, dichlorodiphenyl- (9CI) (8CI)	1135-99-5	$C_{12}H_{10}Cl_2Sn$	dichlorodiphenyltin; diphenylstannyl dichloride; diphenyldichlorotin; diphenyl-tin chloride; dichlorodiphenylstannane
dicyclohexyltin oxide	stannane, dicyclohexyloxo-(9CI) (8CI)	22771-17-1	$C_{12}H_{22}O\ Sn$	
didodecyltin dibromide	stannane, dibromodidodecyl-(9CI)	65264-08-6	$C_{24}H_{50}Br_2Sn$	di-*n*-dodecyltin dibromide
dioctadecyltin dibromide	stannane, dibromodioctadecyl-(9CI)	65264-09-7	$C_{36}H_{74}Br_2Sn$	di-*n*-octadecyltin dibromide
TRISUBSTITUTED COMPOUNDS				
triethyltin bromide	stannane, bromotriethyl- (9CI) (8CI)	2767-54-6	$C_6H_{15}Br\ Sn$	
triethyltin chloride	stannane, chlorotriehyl- (9CI) (8CI)	994-31-0	$C_6H_{15}Cl\ Sn$	chlorotriethylstannane; chlorotriethyltin; triethylstannyl chloride; triethylchloro-stannane; triethylchlorotin
triethyltin iodide	stannane, triethyliodo-(9CI) (8CI)	2943-86-4	$C_6H_{15}I\ Sn$	triethyliodostannane; triethylstannyl iodide; iodotriethylstannane
triethyltin sulfate	4,6-dioxa-5-thia-3,7-distannanonane, 3,3,7,7-tetraethyl- 5,5-dioxide (9CI) stannane, triethylhydroxy-sulfate (2:1) (8CI)	57-52-3	$C_{12}H_{30}O_4S\ Sn_2$	triethylhydroxytin sulfate; bis(triethyltin) sulfate
triethyltin acetate	stannane, (acetyloxy) triethyl-(9CI); stannane, acetoxytriethyl-(8CI)	1907-13-7	$C_8H_{18}O_2Sn$	acetoxytriethylstannane
triethyltin hydroxide	stannane, triethylhydroxy-(9CI) (8CI)	994-32-1	$C_6H_{16}O\ Sn$	triethylstannanol; triethylstannol; triethyl-hydroxystannane; hydroxytriethylstannane
triethylstannylmethyl (1-propynyl) formal	stannane, triethyl(3-methoxy methoxy)-1-propynyl-	17869-84-0	$C_{11}H_{22}O_2Sn$	
triethylstannylphenyl acetylene	stannane, triethyl(phenylethynyl)-(9CI) (8CI)	1015-27-6	$C_{14}H_{20}Sn$	
1-triethylstannyl-3-trimethylsiloxi-1-propyne	silane, trimethyl((3-triethylstannyl)-2-propynyl)oxy)-	4628-88-0	$C_{12}H_{26}O\ SiSn$	

2-trichloro-1-(butine 1'-oxide)-1-(triethyl stannyloxy)ethane	stannane, (1-(3-butinyloxy) -2,2,2-trichloroethoxy)triethyl-	17869-91-9	$C_{12}H_{21}Cl_3O_2Sn$	
trivinyltin chloride	stannane, chlorotriethenyl- (9Cl); stannane, chlorotrivinyl- (8Cl)	10008-90-9	$C_6H_9Cl\ Sn$	chlorotrivinyltin; chlorotrivinylstannane
tributyltin chloride	stannane, tributylchloro- (9Cl) (8Cl)	1461-22-9	$C_{12}H_{27}Cl\ Sn$	tributylchlorotin; chlorotributylstannane; tributylstannyl chloride
tributyltin fluoride	stannane, tributylfluoro- (9Cl) (8Cl)	1983-10-4	$C_{12}H_{27}F\ Sn$	tributylfluorostannane; fluorotributyltin; tri-*n*-butylstannyl fluoride
bis(tributyltin) oxide	hexabutyldistannoxane distannoxane, hexabutyl- (9Cl) (8Cl)	56-35-9	$C_{24}H_{54}O\ Sn_2$	C-Sn-9; BioMeT TBTO; Bultinox; hexabutyl-distannoxane; oxybis(tributyltin); TBTO; 6-oxa-5,7-distannaundecane, 5,5,7,7-tetra-butyl-; Lastanox T; Vikol AF-25; Vikol LO-25; BioMeT 66; oxybis(tributylstannane); BioMeT SRM; Lastanox T20; bis(tri-butylstannyl)oxide; Lastanox Q; Lastanox F; Stannicide A; tributyltin oxide; Myko-lastanox F
tributyltin acetate	stannane, (acetyloxy)tributyl- (9Cl); stannane, acetoxytributyl- (8Cl)	56-36-0	$C_{14}H_{30}O_2Sn$	acetoxytributyltin; tributylacetoxystannane; tri-*n*-butyltin acetate; acetoxytributyl-stannane; tributylstannylacetate
tributyltin linoleate	stannane, tributyl((1-oxo-9,12-octadecadienyl)oxy)-(Z,Z)-(9Cl); stannane, tributyl(linoleoyloxy)- (8Cl)	24124-25-2	$C_{30}H_{58}O_2Sn$	
tributyltin benzoate	stannane, (benzoyloxy)tributyl- (9Cl) (8Cl)	4342-36-3	$C_{19}H_{32}O_2Sn$	tri-*n*-butyltin benzoate; tin, (benzoyloxy) tributyl-
tributyltin salicylate	phenol, 2-[[(tributylstannyl)oxy] carbonyl]- (9Cl) stannane, tributyl(salicyloyloxy)- (8Cl)	4342-30-7	$C_{19}H_{32}O_3Sn$	tin, tributyl(salicyloyloxy)-
tributyltin methacrylate	stannane, tributyl((2-methyl-1-oxo-2-propenyl)oxy)-(9Cl); stannane, tributyl(methacryloyloxy- (8Cl)	2155-70-6	$C_{16}H_{32}O_2Sn$	tributylstannyl methacrylate; tributyl-(methacryloxy)stannane; tin, tributyl-methacrylate; (methacryloyloxy) tributyl-stannane; tributylmethacryloyloxystannane
tributyltin laurate	stannane, tributyl((1-oxo-dodecyl)oxy)-(9Cl) stannane, tributyl(lauroyloxy)- (8Cl)	3090-36-6	$C_{24}H_{50}O_2Sn$	tin, tributyl(lauroyloxy)-; tributyltin laurate; tributyltin dodecanoate
tributyltin oleate	stannane, tributyl((1-oxo-9-octa decenyl)oxy)-, (Z)-(9Cl); stannane, tributyl(oleoyloxy)- (8Cl)	3090-35-5	$C_{30}H_{60}O_2Sn$	tin, tributyl(oleoyloxy)-; N 5117 (Stauffer); N 5117
trihexyltin acetate	stannane, (acetyloxy)trihexyl- (9Cl); stannane, acetoxytrihexyl- (8Cl)	2897-46-3	$C_{20}H_{42}O_2Sn$	tin, acetoxytrihexyl-; acetoxytrihexyltin
tricyclohexyltin hydroxide	stannane, tricyclohexylhydroxy- (9Cl) (8Cl)	13121-70-5	$C_{18}H_{34}O\ Sn$	Plictran; tricyclohexylhydroxystannane; tricyclohexylhydroxytin; Plyctran; M 3180; Cyhexatin; hydroxytricyclohexylstannane; Dowco 213; tricyclohexylstannanol

Name used in text	CAS name	CAS number	Molecular formula	Synonyms
trioctyltin chloride	stannane, chlotrioctyl- (9CI) (8CI)	2587-76-0	$C_{24}H_{51}ClSn$	tin, chlorotrioctyl-; chlorotrioctyltin; tri-*n*-octyltin chloride; chlorotrioctylstannane
triphenyltin chloride	stannane, chlorotriphenyl- (9CI) (8CI)	639-58-7	$C_{18}H_{15}ClSn$	LS 4442; GC 8993; General Chemicals 8993; TPTC; HOE 2872; Brestanol; Fentin chloride; chlorotriphenyltin; chlorotriphenylstannane; triphenylchlorostannane; triphenylchlorotin
triphenyltin hydroxide	stannane, hydroxytriphenyl- (9CI) (8CI)	76-87-9	$C_{18}H_{16}OSn$	hydroxytriphenyltin; hydroxytriphenylstannane; TPTH; triphenylstannanol; K 19 (VAN); Tenhide; Fenolovo; Du-Ter Extra; Erithane; Vancide KS; Dowco 186; ENT 28009; Du-Ter; Fentin hydroxide
triphenyltin acetate	stannane, (acetyloxy)triphenyl- (9CI); stannane, acetyloxytriphenyl- (8CI)	900-95-8	$C_{20}H_{18}O_2Sn$	Brestan; acetoxytriphenylstannane; Batasan; Phentin acetate; Suzu; acetatotriphenylstannane; GC 6936; ENT 25208; Fentin acetate; Lirostanol; tin triphenyl acetate; TPTA; triphenylacetostannane; VP 19-40; Brestan 60; Liromatin
p-bromophenoxy triethyltin	stannane, (*p*-bromophenoxy) triethyl- (8CI)	20961-09-5	$C_{12}H_{19}BrOSn$	
TETRASUBSTITUTED COMPOUNDS				
tetramethyltin	stannane, tetramethyl- (9CI) (8CI)	594-27-4	$C_4H_{12}Sn$	tetramethylstannane
tetraethyltin	stannane, tetraethyl- (9CI) (8CI)	597-64-8	$C_8H_{20}Sn$	tetraethylstannane
tetrabutyltin	stannane, tetrabutyl- (9CI) (8CI)	1461-25-2	$C_{16}H_{36}Sn$	tetra-*n*-butyltin; tetrabutylstannane
tetraisobutyltin	stannane, tetrakis(2-methylpropyl)- (9CI); stannane, tetraisobutyl- (8CI)	3531-43-9	$C_{16}H_{36}Sn$	tetraisobutylstannane
tetraphenyltin	stannane, tetraphenyl- (9CI) (8CI)	595-90-4	$C_{24}H_{20}Sn$	tetraphenylstannane
tetraoctyltin	stannane, tetraoctyl- (9CI) (8CI)	3590-84-9	$C_{32}H_{68}Sn$	tetra-*n*-octyltin; tetra-*n*-octylstannane
stannous octanoate	octanoic acid, tin(2+) salt (9CI) (8CI)	1912-83-0	$C_8H_{16}O_2$ 1/2Sn	tin octanoate; stannous dioctanoate; tin(II)octanoate; stannous caprylate; stannous octoate
tin (II) cyclopentadienyl	stannocene (9CI)	1294-75-3	$C_{10}H_{10}Sn$	tin, di-π-cyclopentadienyl-

TIN AND ORGANOTIN COMPOUNDS: A PRELIMINARY REVIEW

This is the first volume in the UNEP/WHO Environmental Health Criteria series containing a preliminary review of environmental health aspects of a group of chemicals. Such reports are prepared in accordance with the second objective of the WHO Environmental Health Citeria Programme "to identify new or potential pollutants by preparing preliminary reviews on health effects of agents likely to be used in industry, agriculture, in the house, and elsewhere" (WHO, 1976). Organometallic tin compounds are being used in increasing amounts for a variety of applications and the annual world production has risen from less than 50 tonnes in 1950 to about 25 000 tonnes in 1975. One of the main applications is the use of dialkyl- and, to a much lesser extent, monoalkyltin compounds in the stabilization of poly(vinyl chloride). Other applications include the use of tributyltin compounds as industrial biocides and surface disinfectants and the use of triphenyltin and tricylohexyltin compounds as agricultural fungicides and agricultural acaricides.

Preliminary reviews differ from the criteria documents in that they do not contain a separate section on health risk evaluation and that the procedure for their preparation is simpler. Draft preliminary reviews are not submitted for comments to the national focal points for the WHO Environmental Health Criteria Programme. The first draft is reviewed by a Task Group of experts, and on the basis of their comments, a final draft is prepared and scientifically edited by the WHO Secretariat. However, individual members of the Task Group and other experts may be consulted during the scientific editing of the documents.

The first draft of the present document was prepared by Dr L. Fishbein, National Center for Toxicological Research, Jefferson, AR, USA. Dr A. E. Martin, formerly Principal Medical Officer, Department of Health and Social Security, London, England, assisted the Secretariat in the preparation of a revised first draft, which was circulated to the members of the Task Group prior to the meeting. The Task Group on Environmental Health Aspects of Tin and Organotin Compounds met in Geneva from 10—14 March 1975 to review and revise this draft, and, on the basis of their comments, a final draft was prepared by the Secretariat. The Secretariat wishes to acknowledge the most valuable assistance in the final phases of preparation of the document of Dr Renate Kimbrough, Center for Disease Control, Atlanta, GA, USA, Professor Magnus Piscator, Department of Environmental Hygiene, Karolinska Institute, Stockholm, Sweden, Dr Robert J. Horton, US Environmental Protection Agency, Research Triangle Park, NC, and Dr Warren T. Piver, National Institute of Environmental Health Sciences, Research

Triangle Park, NC, USA. The help is also gratefully acknowledged of Dr H. Nordman, Institute of Occupational Health, Helsinki, Finland, who assisted both in the preparation of the final draft and in the final scientific editing of the document and of Professor C. Schlatter and Dr R. Utzinger, Institut de Toxicologie, Ecole Fédérale Polytechnique et Université de Zürich, Dr S. Valley and Dr D. F. Walker, National Library of Medicine, Department of Health, Education and Welfare, USA, and Dr A. Stiles, Consultant, Department of Environmental Health, WHO, Geneva, who helped in compiling the list of organotin compounds.

This document is based primarily on original publications listed in the reference section. However, several publications reviewing the health effects of inorganic and organotin compounds have also been used. These include reviews by Barnes & Stoner (1959), Browning (1969), FAO/WHO (1971), International Labour Office (1972), Kimbrough (1976), MacIntosh (1969), National Institute of Occupational Safety and Health (1976), and Piver (1973).

Details of the WHO Environmental Health Criteria Programme, including definitions of some of the terms used in the documents, may be found in the general introduction to the Environmental Health Criteria Programme, published together with the environmental health criteria document on mercury (Environmental Health Criteria I — Mercury, Geneva, World Health Organization, 1976) and now available as a reprint.

1. SUMMARY AND RECOMMENDATIONS FOR FURTHER RESEARCH

1.1 Chemistry and Uses of Tin Compounds

1.1.1 Inorganic tin

The annual world production of tin is around 225 000 tonnes, about 70% of which is obtained from ores, the remaining 30% being recovered from scrap metal. Tin is mainly used in tinplated containers, but it is also extensively used in solders, in alloys such as bronzes, babbit, pewter, and type metal, and in more specialized alloys such as dental amalgams and the titanium alloys used in aircraft engineering.

Inorganic tin compounds, in which the element may be present in the oxidation states of +2 or +4 are used in a variety of industrial processes for the strengthening of glass, as a base for colours, as catalysts, as stabilizers in perfumes and soaps, and as dental anticariogenic agents.

1.1.2 Organotin compounds

Organotin compounds are classified as R_4Sn, R_3SnX, R_2SnX_2, and $RSnX_3$. In compounds of industrial importance, R is usually a butyl, octyl, or phenyl group and X, a chloride, fluoride, oxide, hydroxide, carboxylate, or thiloate. So far, monosubstituted organotin compounds ($RSnX_3$) have had a very limited application, but they are used as stabilizers in poly(vinyl chloride) films. Disubstituted organotin compounds R_2SnX_2) are mainly used in the plastics industry, particularly as stabilizers in poly(vinyl chloride). They are also used as catalysts in the production of polyurethane foams and in the room-temperature vulcanization of silicones. Trisubstituted organotin compounds (R_3SnX) have biocidal properties that are strongly influenced by the R-groups. The most important of these compounds are the tributyl-, triphenyl-, and tricyclohexyltin compounds, which are used as agricultural and general fungicides, bactericides, antihelminthics, miticides, herbicides, molluscicides, insecticides, nematocides, ovicides, rodent repellents, and antifoulants in boat paints. The tetrasubstituted organotin compounds (R_4Sn) are mainly used as intermediates in the preparation of other organotin compounds.

1.2 Analytical Methods

A wide variety of analytical methods is available for the determination of tin at low concentrations.[a] However, these methods have rarely been compared with regard to their suitability for application to a particular problem and, on the basis of available information, it is not possible to recommend a specific analytical technique for a particular application.

Inorganic tin in food and biological materials is usually determined by atomic absorption. Other spectroscopic methods have also been used with satisfactory accuracy and precision, including emission spectroscopy for air, water, and food samples and neutron activation analysis for geological samples.

Many analytical methods have been used for the determination of organotin compounds. Atomic absorption and other spectroscopic methods combined with chromatography have been used for the

[a] Throughout the document the word concentration indicates mass concentration unless otherwise stated.

estimation of diorganotin compounds. Pesticide residues have been determined by spectroscopic methods and gas-liquid or thin-layer chromatography. However, reliable methods have still to be developed for the quantitative extraction, separation, and determination of many individual tin species in mixtures containing both inorganic tin and organotin compounds that may occur in various media.

1.3 Environmental Concentrations and Exposures

1.3.1 Environmental exposures

On the whole, contamination of the environment by tin is only slight. The levels of pollution arising from gases and fumes, waste slag, and liquid wastes from tin processing are low because of the high degree of recovery and reprocessing used in this industry.

Concentrations of tin in air are often below the detection limits and, when detected, the levels are generally below 0.2—0.3 $\mu g/m^3$, except in the vicinity of industrial sources of emission, where concentrations up to 5 $\mu g/m^3$ may occur.

Tin has not always been found in soils and plants; however, it is possible that in some cases, concentrations have been below the detection limits. Tin concentrations in soils are generally below 200 mg/kg but in regions of tin-containing minerals, the levels may exceed 1000 mg/kg. The small amount of evidence available concerning the uptake of tin by crops suggests that soil concentrations do not markedly influence its uptake by plants.

Tin has been detected only occasionally in river and municipal waters. Values exceeding 1 μg/litre are exceptional, although values as high as 30 μg/litre have been found in drinking water. Sea water concentrations are of the order of 3 μg/litre. Organotin compounds may enter water, for example, from antifouling paints on the bottoms of ships or from molluscicides, which, to be effective, should be present at concentrations of about 1 mg/litre.

Food is the main source of tin for man. A diet composed principally of fresh meat, cereals, and vegetables, is likely to contain a mean tin concentration of about 1 mg/kg. Larger amounts of tin exceeding 100 mg/kg may be found in foods stored in plain cans and, occasionally, in foods stored in lacquered cans. Some foods such as asparagus, tomatoes, fruits, and their juices tend to contain high concentrations of tin if stored in unlaquered cans. Organotin compounds may be introduced into foods through the use of such compounds as pesticides and, to some extent, through migration of tin from poly(vinyl chloride) materials. However, the levels of organotin compounds in food are generally below 2 mg/kg.

Experimental studies have provided evidence of the biotransformation of some triphenyl-, and tricyclohexyltin compounds. There are also limited data suggesting methylation of tin by organisms present in the environment. From the available information, it appears that bioconcentration of tin and organotin compounds of a magnitude that might endanger life or the environment is unlikely to occur.

The estimated mean total daily intake of tin by man ranges from 200 μg to 17 mg. A diet consisting of fresh foods probably provides about 1—4 mg/day. The likely daily intake from water is estimated to be less than 30 μg/day, and the daily amount entering the body from air, less than 1 μg.

1.3.2 Occupational exposure

Several technological operations associated with the processing of tin are known to cause excessive occupational inhalation exposure to tin oxide which may result in a benign pneumoconiosis termed stannosis, many cases of which have been reported in the past.

Workers involved in the processing of di- and trisubstituted organotin compounds may be subject to excessive exposure from time to time. Workers spraying fields or treating plants with trialkyl- or triaryltin compounds may also run the risk of exposure to these compounds.

1.4 Metabolism

1.4.1 Inorganic tin

The extent of absorption through the respiratory route has still to be assessed. The absorption of ingested inorganic tin is likely to be less than 5%, although figures as high as 20% have been suggested. Gastrointestinal absorption is influenced by the oxidation state, tin(II) compounds being more readily absorbed than tin(IV) compounds. The anion complement may also influence the rate of absorption.

Absorbed tin leaves the vascular system rapidly. Bone is the main site of deposition and the highest concentrations of tin are found in the lung, kidney, liver, and bone. Penetration of the blood-brain and placental barriers appears to be very slight. With the exception of the lungs, inorganic tin does not accumulate in organs with increasing age.

Absorbed inorganic tin is mainly excreted in the urine. The fraction excreted with the bile varies with the type of compound and is probably below 15%.

1.4.2 Organotin compounds

In general, organotin compounds are more readily absorbed from the gut than inorganic tin compounds; allowance must be made, however, for the great variations found between different compounds and different species. As a rule, tin compounds with a short alkyl chain are more readily absorbed from the intestinal tract. The trialkyltin compounds are usually well absorbed through the skin. As far as distribution is concerned, the highest concentrations in rats, guineapigs, rabbits, and hamsters have mostly been detected in the liver. Trisubstituted organotin compounds have been found in the brain of various species but the form of tin present in the brain has not been satisfactorily identified.

Many organotin compounds are transformed, to some extent, in the tissues. The dealkylation and dearylation of tetra-, tri-, and disubstituted organotin compounds seem to occur in the liver, but the dealkylation of diethyltin compounds appears to take place both in the gut and in tissues of other organs. The mode of excretion of organotin compounds largely depends on the type of the compound. For example, ethyltin trichloride seems to be mainly excreted with the urine, but diethyltin is eliminated with the faeces, urine, and the bile. Triethyltin is not only excreted with the urine, but, at least in lactating sheep, also with the milk. The route of excretion for many compounds is not known. The biological halftime of different organotin compounds varies and many compounds are slow to disappear from the organs. Usually the biological halftime seems to be longer in the brain than in other organs.

1.5 Effects on Experimental Animals

Although there is evidence that tin is essential for the normal growth of rats, no evidence exists that it is essential for other species including man.

1.5.1 Inorganic tin

1.5.1.1 *Local effects*

Many of the reported effects of inorganic tin are localized because of its irritant properties. Vomiting and diarrhoea are typical signs that follow oral intake of foods with a high tin content. In cats, tin concentrations of 540 mg/litre or 1370 mg/litre in orange juice caused vomiting in 1/11 animals and 3/10 animals, respectively. However, these levels did not produce any effects in dogs. The only adverse effect produced in guineapigs by both short-term and prolonged exposure to 3 mg of tin tetrachloride per m^3 of air was transient irritation of the nose and eyes, but these findings have not been corroborated. Application of 1% tin(II) chloride or 0.25% tin(II) fluoride to the abraded skin of rabbits caused intradermal pustule formation and epidermal destruction, but did not have any effect on intact skin.

1.5.1.2 *Systemic effects*

The major systemic effects of inorganic tin salts in animals include ataxia, twitching of the limbs, and fore-and hindleg weakness progressing to paralysis. In rats, growth retardation and decreased haemoglobin levels may follow administration of tin(II) chloride, orthophosphate, sulfate, oxalate, and tartrate at a dietary level of 3 g/kg. However, administration of iron prevents the development of anaemia. Higher dietary levels of tin (10 g/kg) over several weeks may induce testicular degeneration, pancreatic atrophy, and a spongy state of the white matter of the brain. Doses of pentafluorostannite of 100 mg/kg body weight may also affect growth, and a dose-related decrease in haemoglobin levels may be seen with doses exceeding 100 mg/kg; no effect on growth was found at a dose of 20 mg/kg administered orally to rats. A single intravenous injection of pentafluorostannite at a concentration of 35 mg/kg body weight or tin(II) chloride dihydrate ($SnCl_2$-$2H_2O$) at 44.4 mg/kg in rats produced extensive necrosis, mainly in the proximal tubules of the kidney. A subcutaneous dose of tin(II) chloride at a concentration of 5.6 mg/kg body weight caused a 20—30 fold increase in the haem oxidation activity in the kidney; this effect was dose-related. Administration of tin(II) chloride at a concentration of 5 mg/litre, from weaning to natural death, did not affect longevity in mice or in male rats, but caused a decrease in longevity in female rats combined with an increased incidence of fatty degeneration of the liver. There is no conclusive evidence concerning the carcinogenicity or teratogenicity of inorganic tin.

1.5.2 Organotin compounds

1.5.2.1 *Local effects*

Some butyltin compounds are known to produce gastrointestinal irritation; submucosal, subserosal, and intraluminar haemorrhages were seen in mice after a single oral dose of 4000 mg/kg body weight. Dibutyltin dichloride administered at a dose of 50 mg/kg body weight per day, for one week, produced gastroenteritis in rats. Gastroenteritis was also produced in rats by administration of tricyclohexyltin hydroxide (25 mg/kg body weight per day, for 19 days).

Dermal application of dibutyltin dichloride (10 mg/kg body weight per day, for 12 days) caused severe local damage. Local irritation was produced in rats by applications to the shaved skin of bis(tributyltin) oxide in doses of 0.36—0.95 mg/kg; necrosis was produced at doses of 1.4—185 mg/kg. Triphenyltin acetate also irritated the skin of the rat, whereas triphenyltin hydroxide was reported not to irritate the skin of the rabbit but to be extremely irritating to the eyes.

1.5.2.2 *Systemic effects*

The systemic effects of monosubstituted, disubstituted, and trisubstituted organotin compounds differ. In general, mono- and diorganotin compounds are less toxic than triorganotin compounds. The toxicity of trialkyltin compounds decreases as the number of carbon atoms in the alkyl chain increases.

Dibutyltin compounds can produce inflammatory changes in the bile duct. Single oral doses of dibutyltin dichloride at 50 mg/kg body weight produced this effect in rats, and higher doses produced more severe injury; necrotic changes were also produced in the liver of mice and rats. Bile duct injury in rats and rabbits was seen following dermal application of dibutyltin dichloride (10 mg/kg body weight). Dioctyltin compounds produced slight changes in the germinal centres of the spleen and steatosis of hepatocytes in mice at a single oral dose of 4000 mg/kg body weight. Pulmonary oedema may be seen in rats following intravenous administration of diethyl-, dipropyl-, diisopropyl-, and dipentyltin compounds. Dibutyltin compounds can slow down growth in rats. The no-observed-effect dietary level was reported to be 40 mg/kg for a 3 month feeding period and 20 mg/kg for 6 months. Recent studies showed that dioctyltin dichloride and dibutyltin dichloride administered at dietary levels of 50 and 150 mg/kg, respectively for 6 weeks, caused a dose-dependent atrophy of the thymus and thymus-dependent organs and suppression of the immunological response in rats, but not in mice and guineapigs.

Some trialkyltin compounds produce a characteristic lesion in the central nervous system consisting of oedema throughout the white matter. Orally administered trimethyl- and triethyltin compounds are more potent in inducing this lesion than the higher homologues. The first changes in the rat brain were visible after 3 days of administration of triethyltin hydroxide at a dietary level of 20 mg/kg. Maximum changes were found after 2 weeks. Typical signs of such intoxication included prostration and weakness of the limbs progressing to flaccid paralysis. The effects disappeared when exposure ceased.

Administration of triphenyltin compounds produced a reduction in weight and in food intake in many species. Lethargy was a typical symptom and histological changes in the liver and spleen were also seen. A decreased immunological response with a reduction in the number of leukocytes and of plasma cells in the lymph nodes of guineapigs has been reported. A 2-year study indicated a no-observed-effect level for triphenyltin acetate of 0.1 mg/kg body weight per day.

A single intrarumenal dose of tricyclohexyltin hydroxide at 50 mg/kg body weight produced central nervous depression and diarrhoea in sheep, whereas a dose of 15 mg/kg did not result in any adverse effects. At higher doses, pulmonary congestion, tracheal haemorrhage, enteritis, and electrocardiographic changes were seen. No-observed-effect doses for long-term intake in the rat and dog were given as 3 mg/kg body weight per day and 0.75 (mg/kg) per day, respectively.

Tetraalkyltin compounds may produce muscular weakness, paralysis, respiratory failure, tremors, and hyperexcitability as acute effects in mice and dogs, while late effects are similar to those seen with triorganotin poisoning.

There is no evidence that organotin compounds are carcinogenic or teratogenic. Reported effects of triphenyltin hydroxide on the testes and ovaries of rats require further confirmation.

Information concerning the mechanism of the toxic action of organotin compounds is inadequate. Several dimethyl- and dioctyltin compounds inhibit the oxidation of keto-acids and block mitochondrial respiration. Trialkyltin compounds inhibit oxidative phosphorylation.

1.6 Effects in Man

1.6.1 Inorganic tin

Inhalation of elemental tin does not produce any effects in man, whereas extended exposure to tin(IV) oxide dust and fumes can produce a benign pneumoconiosis termed stannosis. This condition develops after at least 3—5 years of exposure and is characterized by small dense shadows in the pulmonary X-ray picture without impairment of pulmonary function. Fibrosis is not seen. The generally-accepted maximum allowable concentration of tin(IV) oxide in the air of work rooms of 2 mg/m^3 appears to give protection against this disorder.

Symptoms that have been reported following ingestion of food with a high tin content include nausea, vomiting, diarrhoea, stomach cramps, fatigue, and headache. The lowest concentration of tin reported in association with such outbreaks was about 250 mg/litre in canned orange and apple juice. Five human volunteers did not experience any symptoms from the ingestion of fruit juices containing concentrations of 500—730 mg/litre but all had gastrointestinal disturbances at a level of 1370 mg/litre (corresponding to 4.4—6.7 mg/kg body weight). Ingestion of 50 mg of tin through eating canned peaches that contained tin concentrations of about 300—600 mg/kg caused acute symptoms in 2 out of 7 persons. The relative importance of, on one hand, the total amount of tin ingested and, on the other hand, the concentration of tin in relation to the development of symptoms has not been satisfactorily assessed.

1.6.2 Organotin compounds

1.6.2.1 *Local effects*

Dibutyl- and tributyltin compounds produced skin irritation in workers 1—8 h after contact. Experimental application to the skin of volunteers showed that some compounds (e.g., dibutyltin dichloride and tributyltin chloride) produced this effect, whereas others such as dibutyltin maleate and tetrabutyltin did not. Di- and tributyltin compounds caused eye irritation after brief contact. A 20% solution of triphenyltin acetate produced irritation of the skin and the mucous membranes of the upper respiratory tract while tricyclohexyltin hydroxide was reported not to cause skin irritation at a concentration of 0.01 mg/kg body weight.

1.6.2.2 *Systemic effects*

The majority of accidental poisonings involving systemic effects have been due to occupational exposure to triphenyltin acetate. Systemic effects reported to have followed both dermal and inhalation exposure include general malaise, nausea, gastric pain, dryness of the mouth, vision disturbance, and shortness of breath. Hepatomegaly and elevated levels of liver transaminase activity have been found in some cases. Recovery has generally been complete but liver damage has been known to persist for up to 2 years.

The hazard associated with the use of organotin compounds was unmasked by an episode of intoxication in 1954 involving over 200 cases, 100 of which were fatal. The cause was the ingestion of an oral preparation containing diethyltin diiodide at 15 mg/capsule. It was suggested, however, that ethyltin triiodide, triethyltin iodide, and tetraethyltin were present as impurities. Predominant symptoms and signs included severe headache, nausea and vomiting, visual and psychological disturbances, and sometimes loss of consciousness. At autopsies and decompressive surgery, cerebral oedema of the white matter was found. In many cases, symptoms lasted for at least 4 years; follow-up information on the subjects involved is not available.

1.7 Recommendations for Further Studies

1.7.1 Analytical methods

More information is needed on the specificity, precision, and accuracy of methods for the determination of inorganic tin compounds. Data concerning interlaboratory comparisons of the methods used are also lacking. In view of the variable results obtained in studies on the tin contents of various materials and tissues, the use of reference laboratories is recommended. Better methods are needed for the quantitative extraction and separation of the various organotin compounds present in environmental and biological samples. As organotin compounds used as pesticides or stabilizers occur in foods in minute amounts only, more sensitive methods for their measurement are needed.

1.7.2 Environmental data

More information regarding bioconcentration is needed. The fate of organotin compounds entering water is largely unknown. The possibility of the methylation of tin by organisms present in the environment is of particular interest.

A wide variety of results concerning the daily intake of tin has been reported. Although a certain range is to be expected, further investigations of the concentrations of tin in food and water, and of dietary intake are needed.

1.7.3 Metabolism

Information on the rate of absorption of tin from the gastrointestinal tract is insufficient and little is known about the absorption of various compounds through the respiratory tract which may be of importance in occupational exposure. Furthermore, there is a gap in information concerning the rate of absorption of organotin compounds through the skin.

Many of the studies conducted on the distribution of tin in human tissues may be unreliable because of the analytical methods available at that time: thus, data on tissue contents should be obtained using as sensitive methods as possible with emphasis on the precision, specificity, and accuracy of the assays employed. Information on tin concentrations in newborn infants compared with adults is lacking, and data concerning tin concentrations in the tissues of occupationally-exposed persons compared with unexposed populations are not available. The increasing development and use of new organotin compounds will necessitate futher studies on the metabolism of such compounds.

At present, there is an obvious lack of information on the biotransformation of several organotin compounds. Data concerning the accumulation and retention times of various compounds in animal tissues are also desirable. Finally, the route or routes of excretion of many organotin compounds are completely unknown.

1.7.4 Effects

Probably the most conspicuous lack of information concerns the mechanism of action of various organotin compounds. More information should be obtained on the carcinogenicity, teratogenicity, and mutagenicity of these compounds. Compounds used industrially should be studied with reference to possible allergenic properties. Recently reported results suggest that the effects of various organotin derivatives on the immune system should be studied in more detail. Moreover, properly conducted studies on the effects on the sex glands of different species using multigeneration studies seem urgent. The importance of longitudinal epidemiological studies on occupationally-exposed populations and the usefulness of information that may be obtained through follow-up studies of accidental intoxication should be emphasized.

2. CHEMISTRY AND ANALYTICAL METHODS

Tin can form a variety of both inorganic and organometallic compounds. These two classes of compounds have different chemical and physical properties which make them suitable for different applications in industry, agriculture and elsewhere. They also have different toxicities and require separate assessments of health risk. The inorganic chemistry of tin has been described in standard texts on inorganic chemistry such as those by Cotton & Wilkinson (1972, 1976) and Heslop & Jones (1976). Sources of information on new developments in the organometallic chemistry of tin include a review by van der Verk (1972) and a collection of papers presented at a symposium of the American Chemical Society (Zuckerman, 1976).

2.1 Elemental Tin

Tin (atomic number 50; relative atomic mass 118.70) is an element of group IVb of the periodic system, together with carbon, silicon, germanium, and lead. It exists in three allotropic modifications. At room temperature, the stable form is a metallic form called β- or white tin. White tin is a silver-white, lustrous and soft metal with considerable ductility, and can be rolled into very thin "tin foil". Its density is 7.27, melting point 231.9°C, and boiling point 2507°C. At ordinary temperatures, it is stable in both air and water. Below 13.3°C, the stable form of tin is the non-metallic grey tin (α-tin). Above 161°C, the stable modification is the so-called brittle tin, another metallic modification.

Metallic tin is normally covered with a thin protective film of tin dioxide. Because tin is resistant to cold acids, a cohesive tin layer will protect iron from corrosion (tin plate). However, if the layer is damaged, the iron will rapidly corrode. Tin-plate used in the food industry should not contain lead, not only because lead is toxic, but also because its presence aids the corrosion of tin by dilute organic acids.

Neutral aqueous salt solutions react slowly with metallic tin in the presence of oxygen but solutions containing nitrates, iron(II) chloride or sulfate, aluminium chloride or tin(IV) chloride dissolve elemental tin.

Tin can form inorganic compounds in the oxidation state +2 (Sn^{II}, tin(II) compounds, or stannous compounds), and in the oxidation state +4 (Sn^{IV}, tin(IV) compounds, or stannic compounds). Because of their different physicochemical properties, it is useful to discuss the 2 groups of compounds separately.

2.2 Tin(II) Compounds

Tin(II) compounds are generally more ionic than tin(IV) compounds. They are unstable in dilute aqueous solutions, are easily oxidized, and normally contain some Sn^{IV}; after some time, hydrolysis occurs with the formation of the hydrated tin(II) oxide ion $[Sn_3(OH)_4]^{2+}$.

Tin(II) chloride is readily soluble in small amounts of water. It is a fairly good reducing agent, and has many uses in industry, particularly as a mordant in dye printing. Aqueous solutions of tin(II) chloride become turbid on dilution because a basic salt is precipitated. The fluoride (SnF_2) is slightly soluble in water. It is used in fluoride-containing toothpastes. In aqueous solutions, SnF_3^- is the major ion but other ions such as SnF^+ and $Sn_3F_5^-$ are also present. Tin(II) sulfate is a good source of Sn^{II}. Its solubility decreases with temperature.

Tin(II) oxide (SnO) is a stable, blue-black crystalline solid. It reacts with both mineral and organic acids, and dissolves in sodium hydroxide solutions forming stannites, which probably contain the SnO_2^{2-} ion.

Other tin(II) compounds that have practical applications are tin(II) acetate, tin(II) arsenate, tin(II) fluoroborate, tin(II) pyrophosphate, and several tin(II) salts or organic acids, such as tin(II) oxalate and tin(II)-2-ethylhexoate (tin(II)"octoate").

2.3 Tin(IV) Compounds

Tin in the oxidation state + 4 forms a large number of inorganic compounds as well as organometallic compounds, which are discussed in section 2.4. Some tin(IV) compounds, such as tin(IV) oxide (SnO_2), have long been used in industry; others, e.g., tin(IV) chloride ($SnCl_4$), have found technological applications more recently. Also of practical importance are the stannates, compounds in which the tin atom is part of an anion. The structure of stannates can be represented by $M_nSn(OH)_6$, where M is a metal ion.

The physical properties of tin tetrahalides, except those of SnF_4, correspond to the properties of covalent halides of carbon and silicon. Tin(IV) chloride is a colourless liquid that fumes in moist air and becomes turbid because of hydrolysis when complex ions such as $[SnCl_3(OH)_3]^{2-}$ are formed. The addition of a limited amount, of water to tin(IV) chloride results in the formation of a crystalline hydrate, $SnCl_4.5H_2O$, the ionic character of which is probably due to the presence of a complex ion $[Sn(H_2O)_4]^{4+}$.

Tin(IV) oxide occurs naturally as the mineral cassiterite. It has a very high melting point (1127 °C) and has wide application in industry. The fusion of tin(IV) oxide with sodium or potassium hydroxide yields stannates.

Other tin(IV) compounds that have found practical applications include tin(IV) sulfide, tin(IV) vanadate, and tin(IV) molybdate.

2.4 Organometallic Compounds of Tin

Organometallic tin compounds or organotin compounds have one or more carbon-tin covalent bonds that are responsible for the specific properties of such molecules. Essentially all organometallic tin compounds are of the Sn^{IV} type. The only well established compound with tin in the oxidation state +2 is the tin(II)cyclopentadienyl, $C_{10}H_{10}Sn$. There are four series of organotin compounds depending on the number of carbon-tin bonds. These series are designated as mono-, di-, tri-, and tetraorganotin compounds with the general structure:

$$R_nSn\ X_{4-n}$$

where R = an alkyl or aryl group
Sn = the central tin atom in the oxidation state +4
X = a singly charged anion or an anionic organic group.

In the organotin compounds of practical importance, R is usually a butyl, octyl, or phenyl group and X is commonly chloride, fluoride, oxide, hydroxide, carboxylate, or thiolate.

Monoorganotin compounds, $RSnX_3$, are known but so far have found only limited application, for example, butyltin sulfide is used as a stabilizer in poly(vinyl chloride) (PVC) film.

Diorganotin compounds, R_2SnX_2, are chemically reactive and most of their applications are based on this property. They are used as stabilizers of PVC, as catalysts in the production of polyurethane foams, and in the cold-curing of silicon elastomers.

Triorganotin compounds, R_3SnX, are the most important class of organotin chemicals. They are biologically very active and are widely used as biocides. The chemical nature of the R group has a strong influence on the biological properties of these compounds. The X-group, on the other hand, influences their solubility and volatility. The two most important groups of triorganotin compounds are tributyltin and triphenyltin derivatives.

Tetraalkyl- and tetraaryltin compounds are primarily used as intermediates in the preparation of other organotin compounds. Tetraalkyltin compounds are colourless and the compounds of lower molecular weight are liquids at room temperature. The tetraaryltin compounds are solids. Tetraorganotin compounds possess typical covalent bonds and are stable in the presence of air and water. Tetrabutyltin, $Sn(C_4H_9)_4$ is a colourless oily liquid with a distinct odour. Tetraphenyltin, $Sn(C_6H_5)_4$, is a white crystalline powder, soluble in organic solvents and insoluble in water.

Since 1974, a new class of organotin compounds, called estertins, has been developed for use as stabilizers in poly(vinyl chloride). Their general structure is $(R\text{-}O\text{-}CO\text{-}CH_2\text{-}CH_2)_2SnX_2$ or $R\text{-}O\text{-}CO\text{-}CH_2\text{-}CH_2SnX_3$ where X may be, for example, isooctylmercaptoacetate. They have a comparatively low volatility and extractability (Lanigon & Weinberg, 1976).

Solubility data for organotin compounds are incomplete. In general, their solubility in water at ambient temperatures is of the order of 5 to 50 mg/litre, but they are very soluble in many common organic solvents, such as alcohol, ethers, and halogenated hydrocarbons.

Commercial products are usually pure chemicals since, for technological reasons, scrupulous care must be taken to avoid metal contamination during manufacture. The impurities are primarily solvent residues remaining from the product purification and separation processes.

The carbon-tin bond is susceptible to nucleophilic and electrophilic attack, e.g., hydrolysis, solvolysis, acidic and basic attack, and halogenation. Water has little effect on symmetrical saturated organotin compounds. Dialkyltin compounds react spontaneously with moisture and air to form dialkyl hydrated oxides. Photochemical reactions of organotin compounds are mentioned in connexion with environmental transport and transformations (section 4). The physicochemical properties of some organotin compounds have been listed by Weast (1976).

2.5 Analytical Methods

2.5.1 Determination of inorganic tin

2.5.1.1 *Atomic absorption spectroscopy*

Atomic absorption spectroscopy is the method most widely used for the detection of low concentrations of tin. In general, the lowest limit of detection is obtained with a fuel-rich, air-hydrogen flame, e.g., about 1—1.5 mg/kg compared with 2—2.5, and 4—5 mg/kg for air-ethylene and nitrous oxide-ethylene flames, respectively (Christian & Feldman, 1970). The detection limits at the 3 absorption lines, 2246.1, 2354.8, and 2863.3 nm do not differ much. Using some of the procedures mentioned later, the detection limit may be reduced to about 0.1—0.5 mg/kg.

Several atomic absorption techniques, modified to suit specific purposes, have been reported. A detection limit of 0.5 μg/litre was obtained when a hydride generation technique using sodium borohydride and a flame-heated silica atomizing tube was used in the air-acetylene atomic absorption determination of tin in solution (Thompson & Thomerson, 1974). Capacho-Delgado & Manning (1966), using a high intensity hollow cathode lamp as the source, reported a detection limit of 0.1 mg/litre for water solutions of metallurgical samples, while Schallis & Kann (1968) determined tin in lubricating oils with a detection limit of about 0.5 mg/litre. Carbon filament atomic absorption spectroscopy was employed by Everett et al. (1974) to determine tin(II) chloride in aqueous and xylene solutions and tin octoate in oil solution.

Atomic absorption spectroscopy has been used extensively for the determination of tin in foods (Allan, 1962; Amos & Willis, 1966; Capacho-Delgado & Manning, 1966; Christian & Feldman, 1970; Gatehouse & Willis, 1961) and particularly in canned foods, including fruit juice, fruits, and vegetables (Catala et al., 1971; Price & Roos, 1969; Sato et al., 1973; Shiraishi et al., 1972; Woidich & Pfannhäuser, 1973). A detection limit of 0.5 mg/kg has been reported for the determination of tin in canned fruit juice using a nitrous oxide-acetylene flame (Price & Roos, 1969).

Engberg (1973) compared atomic absorption with a spectrophotometric method using 2-(3,4,-dihydroxyphenyl)-3,5,7-trihydroxy-4H-1-benzopyran-4-one (quercetin) for the determination of tin in food. The methods gave similar results at concentrations of tin generally found in canned food, but at very low concentrations (e.g., organotin residues), the quercetin method was more suitable because of its lower detection limit (section 2.5.1.5).

Atomic absorption spectroscopy has also been used for the determination of tin in biological samples (Pearlman et al., 1970).

2.5.1.2 *Emission spectroscopy*

Emission spectroscopy is a rapid and specific method that has frequently been used for the simultaneous determination of several elements. Unfortunately, this method is exacting, demanding highly qualified personnel, and the cost of the instrument is high. It has been used for the determination of tin in atmospheric samples (Hasegawa & Sugimae, 1971; Keenan & Byers, 1952; Laamanen et al., 1971; Lee et al., 1972; Schroll & Krachsberger, 1970; Sugimae, 1974; Tabor & Warren, 1958), tin in water (Ghafouri, 1970; Konovalov & Kolesnikova, 1969; Rittenhouse et al., 1969), and for tin in various foods including meats (Krylova & Balabuh, 1970), fruits, and vegetables (Chisaka et al., 1973). A sensitivity of 0.04 mg/kg fresh weight was reported by Tihonova & Zore (1968) for vegetables and berries. Several investigators have used emission spectroscopy for the determination of tin in biological samples (Avtandilov, 1967; Geldmacher-von Mallinckrodt & Pooth, 1969; Kas'yanenko & Kul' skaya, 1969; Kehoe et al., 1940; Mulay et al., 1971; Saito & Endo, 1970 Tipton et al., 1963).

2.5.1.3 *Neutron activation analysis*

Although the limit of detection of tin by neutron activation is comparatively high, the technique has been used to determine tin in air samples (Bogen, 1973; Tuttle et al., 1970). It is also the most commonly used method for the determination of tin in geological samples (soils, sediments, rocks, oil) (Johansen & Steinnes, 1969; Obrusnik, 1969; Schramel et al., 1973). A neutron activation electron probe was used by Kurosaki & Fusayama (1973) for the estimation of tin in teeth. Disadvantages of neutron activation include the need for a nuclear reactor and the possibility of interference from formation of other isotopes.

2.5.1.4 *X-ray fluorescence*

X-ray fluorescence, a non-destructive method for the determination of tin, has been used in the analysis of air for elements ranging from titanium to caesium with a detection limit of 0.5 $\mu g/m^3$ of air (Dittrich & Cothern, 1971) and for river water with a detection limit of 20—30 μg/litre for metals in the suspended or particulate form, and 0.25—0.4 mg/litre for ionic metals (Blasius et al., 1972).

2.5.1.5 *Miscellaneous analytical methods*

Among the spectrophotometric methods reported, Kirk & Pocklington (1969) recommended the tin-quercetin method for the estimation of the tin contents of foods at concentrations ranging from 10 to over 500 mg/kg. The smallest absolute amount of tin detectable with the quercetin method has been reported to be about 1 μg (Engberg, 1973). The pyrocatechol violet method has been used for the direct determination of tin(II) and tin(IV) compounds at levels of about 5—50 mg/kg in fats and oils (Lowry & Tinsley, 1972), for the determination of tin in metals, and for the determination of tin concentrations ranging from 0.01—1.0 mg/kg in biological samples (Corbin, 1973). Other spectrophotometric methods have been developed for the determination of tin in food samples using phenylfluorone (Bennet & Smith, 1959; Bergner & Rüdt, 1968; Luke, 1956; Nakamura & Kamiwada, 1973; Smith, 1970) dithiol (Ljaskovskaja & Krasilnikova, 1961), salicyl-idenamino-2-thiophenol (Horio & Nakaseko, 1972) and stilbazo (Kobayashi & Yada, 1968).

Infrared internal reflection spectroscopy was used in studies on tooth enamel treated with tin(II) fluoride (SnF_2) (Krutchkoff et al., 1972).

A fluorimetric method using 3′4′7-trihydroxyflavone with a detection limit of 0.007 μg was reported for the determination of tin in rock, soil, and biological materials (Filer, 1971). Other fluorimetric procedures have been reported using flavonal (Coyle & White, 1957), oxine-5-sulfonic acid (Pal & Ryan, 1956), and the ammonium salt of 6-nitro-2-naphthylamine-8-sulfonic acid (Anderson & Lowy, 1956).

Electroanalytical methods have also been employed for the determination of tin in canned foods. Polarographic analysis was used for the analysis of canned meat (Janitz, 1971; Jovanović et al., 1967), fruits (Miki & Fukui, 1971), and juices (Hayashi, 1969), and for the estimation of tin in cans (Gruenwedel & Patnaik, 1973). Biston et al. (1972) obtained good agreement between polarographic and spectrophotometric (dithiol and phenylfluorone) methods for estimating tin in vegetables at concentrations of 20—400 mg/kg. Anodic stripping voltametry has been used for the determination of tin in water (Portretnyj et al., 1973).

Tin estimations were made in biological samples by means of spark-source mass spectroscopy (Evans & Morrison, 1968; Hamilton et al., 1972/1973) and in canned vegetables and juices by titrimetric analysis (Zohm, 1972).

2.5.2 Determination of organotin compounds

A number of techniques and procedures for the separation of mono-, di-, and trisubstituted organotin compounds with their subsequent determination have been described (Brinkman et al., 1977; Freitag & Bock, 1974; Getzendaner & Corbin, 1972; Kumpulainen & Koivistoinen 1977; Meinema et al. 1978; Soderquist & Crosby, 1978; Woggon & Jehle, 1973, 1975). However, reliable methods have still to be developed for the quantitative extraction and determination of many individual tin species in mixtures containing inorganic tin (IV) and organotin compounds that occur in various media including biological materials, industrial effluents, and river sediments.

2.5.2.1 *Diorganotin compounds*

Compounded poly(vinyl chloride) formulations usually contain 1—2 % of dialkyltin compounds as stabilizers and various techniques have been used for the determination of these organotin stabilizers in foods. Organotin compounds were separated from inorganic tin chromotographically and the inorganic tin was then determined spectroscopically as its catechol violet complex. Koch & Figge (1971) determined the extent of migration of dioctyltin dichloride and dioctyltin bis(2-ethyldexylmercaptoacetate) from PVC bottles into beer by the same method. The catechol violet complex was also used, by Adamson (1962) and by Ross & White (1961), for the determination of dialkyltin compounds in fats and olive oil. Spectroscopic methods involving dithizone (Aldridge & Cremer, 1957; Chapman et al., 1959), diphenylcarbazone (Skeel & Bricker, 1961) and 4-(2-pyridylazo)-resorcinol (Sawyer, 1967) have been described. Dithizone and diphenylcarbazone were used for the determination of diethyl- and dibutyltin compounds respectively, but some difficulties may arise using either of these agents because of their inherent instability in solution. Concentrations of diethyl- and triethyltin compounds ranging up to 30 μg and 20 μg, respectively, could be measured using dithizone (Aldridge & Cremer, 1957). The use of atomic absorption spectroscopy for the determination of dibutyltin dilaurate in animal feeds was described by George et al. (1973).

Neubert (1964) used thin-layer chromatography for the determination of dialkyltin stabilizers obtaining a detection limit of 1 μg of organotin. Thin-layer chromatographic determination of organotin stabilizers using their quercetin chelates yielded a detection limit of 1.3 μg (Wieczorek, 1969). Udris (1971) described a number of schemes for the analysis of commercial tin stabilizers commonly used in poly(vinyl chloride) production. Methods were given for the chemical breakdown of a sample and the subsequent separation and identification of the degradation products. Schemes for the analysis of dialkyltin thio-compounds and dialkyltin carboxylates and hemiesters, respectively, were also presented.

2.5.2.2 *Triorganotin compounds*

A number of procedures have been used for the analysis of organotin fungicide and miticide residues in food, including spectrophotometry (Corbin, 1970; Getzendaner & Corbin, 1972; Trombette & Maini, 1970), gas-liquid chromatography (Gauer et al., 1974), and thin-layer chromatography (Wieczorek, 1969).

Corbin (1970) described a dithiol spectrophotometric method for the determination of trace amounts of tin residues on fruits previously treated with a miticide containing tricyclohexyltin hydroxide as the active component. The detection limit with this method was at least 0.2 kg tin at a concentration of 3 μg/kg. The extraction method was tested for compatibility with 35 elements and only arsenic and antimony seemed likely to interfere.

A sensitive fluorometric technique has recently been developed for the determination of triphenyltin residues in potato samples (Vernon, 1974). Triphenyltin acetate deposits on potato leaves have also been determined polarographically by Coussement (1972). The smallest quantity of organotin fungicide detected by this method was 0.32 $\mu g/cm^2$ leaf.

Freitag & Bock (1974) reported some methods for the extraction of tri-, di-, and phenyltin compounds from mixtures containing these compounds as well as inorganic tin(IV). The separated compounds were determined by radiometric or photometric methods. Several thin-layer chromatographic methods were also described.

Bönig & Heigener (1972) determined the tin contents of plants treated with organotin fungicides by photometric estimation of the phenylfluorone complex. In an alternative method, tin was extracted with quercetin and the tin-quercetin complex determined spectroscopically (Engberg, 1973). Akagi and his collaborators (1972) separated some butyl- and phenylorganotin compounds in vinegar and tomatoes by thin-layer chromatography.

A variety of trialkyl and triaryltin biocides have been determined by a nonaqueous atomic absorption assay (Freeland & Hoskinson, 1970). The limit of detection of this method appeared to be 2—12 mg/litre at 1% absorption. Studies have been reported by

Woggon & Jehle (1973; 1975) and Woggon et al., (1972) in which anodic stripping was used for the determination of a number of alkyl- and aryltin fungicides and their degradation and decomposition products. The minimum detectable amount was about 3.5×10^{-7} moles per litre for all the compounds studied.

Polarography (Kočkin et al., 1969; Tjurin & Flerov, 1970; Tjurin et al., 1969) and oscillopolarography (Geyer & Rotermund, 1969; Shono & Matsumura, 1970) have also been used for the determination of a variety of organotin compounds.

Cenci & Cremonini (1969) described the thin-layer chromatographic determination of 2 commercial organotin pesticides containing triphenyltin acetate and triphenyltin hydroxide, respectively, and their degradation products in various soils.

A gas-liquid chromatographic method was developed by Tonge (1965) for the analysis of butyl-, octyl-, and phenyltin halides. A variety of gas-liquid chromatographic procedures for the determination of a large number of other organotin compounds has been reported (Devjatyh et al., 1968; Dressler et al., 1971, 1975; Geissler & Kriegsmann 1964, 1965).

Gauer et al. (1974) reported a gas-liquid chromatographic method for separating and determining tricyclohexyltin hydroxide and dicyclohexyltin compounds formed by degradation (as the bromide) on strawberries, apples, and grapes treated with a miticide. The practical minimum limits of detection for tricyclohexyltin hydroxide and dicyclohexyltin oxide were 0.1 and 1.0 mg/kg, respectively, in the 3 crops studied.

The quantitative determination of mono-, di-, tri- and tetraalkyltin compounds by gas-liquid chromatography after alkylation was reported by Neubert & Wirth (1975); the same technique was applied for the quantitative detection of tri- and dibutyltin species in dilute aqueous solution by Neubert & Andreas (1976). Application of a liquid-chromatograph coupled with a flameless atomic absorption detector for speciation of trace amounts of triphenyl- and trialkyltin compounds in aqueous solution has been reported by Brinkman et al. (1977). Recently, Meinema et al. (1978) developed a combined gas chromatography/mass spectrometry detection procedure for the quantitative determination of trace amounts of tri-, di-, and butyltin compunds in aqueous solutions.

3. SOURCES OF ENVIRONMENTAL POLLUTION

3.1 Natural Occurrence

Tin is not uniformly distributed over the earth's surface (Goldschmidt, 1958; Schroeder et al., 1964) and, hence, it is not found consistently in plants and soils (Schroeder et al., 1964). Most samples of rock contain tin concentrations of approximately 2—50 mg/kg (Johansen & Steinnes, 1969; Mason, 1966; Onishi & Sandell, 1957; Vinogradov, 1956), although levels of about 260 and 1200 mg/kg have been reported in Czechoslovakian and Norwegian samples of mica (Johansen & Steinnes, 1969).

Of the 9 different tin-bearing minerals found in the earth's crust, only cassiterite (tinstone, tin(IV) oxide) is of major commerical importance (Heindl, 1970), although small quantities of tin are recovered from the complex sulfides, e.g., stannite ($Cu_2S.FeS.SnS_2$); teallite ($PbSnS_2$); cylindrite ($PbSn_4FeSb_2S_{14}$) and canfieldite (Ag_8SnS_6). Over 80% of the world's tin ore occurs in low-grade deposits averaging about 240g of metallic tin per cubic metre.

3.2 Industrial Production

The total world production of tin in 1975 was 236 000 tonnes, about 92% of which was produced as primary tin. Six countries together produced 72% of the total world production, i.e., China (10%, estimated figure), Indonesia (8%), Malaysia (35%), Thailand (7%), United Kingdom (6%), and the USSR (6%, estimated figure). The world production of secondary tin was about 20 000 tonnes, almost 50% of which was produced by France (United Nations, 1977).

The metallurgy of tin is simple, but its extraction from the ore is complicated by the presence of reduced iron that forms "hard head" with the tin, and by the high tin content of the slag produced. Thus, smelting is carried out in 3 stages: (*a*) primary smelting in either reverberatory or blast furnaces; (*b*) retreatment of slags, hard head, and refinery dross; and (*c*) refining of the metallic tin to remove the last traces of impurities. It should be noted that, because of the high cost of tin, dust-collection equipment is necessary for successful operation. Low fume production is one of the advantages of electric smelting (MacIntosh, 1969).

During the last 10 years, the amount of tin recovered from secondary sources has been practically constant (United Nations, 1977). The largest sources of scrap are: clean tin plate clippings from container manufacture; solder in the form of dross or sweepings; dross from tinning pots; sludges from tinning lines; bronze rejects and used parts; babbitt from discarded bearings; and type metal scrap (MacIntosh, 1969). Small quantities of tin are recovered at detinning plants from used tin containers. (Heindl, 1970).

3.3 Tin Consumption

The USA is by far the largest consumer of tin, with Japan, the United Kingdom, the Federal Republic of Germany, and France following in that order. It has been estimated that, in the future, 30% of the total demand for tin will be met by secondary recovery of the metal. The total demand for primary tin from 1968 to the year 2000 has been estimated to lie between 8.5 and 6.2 million tonnes, with a median estimate of 7.5 million tonnes. The world reserve total is approximately 6.5 million tonnes and it is considered likely that new discoveries and increases in known reserves could result in sufficient new tin to meet the median estimate for this period (Heindl, 1970). Recently, nearly 40% of the total primary and secondary tin consumption in the USA has been used in the production of tin-plate, 25% in solders, and 20% in bronze and brass, while smaller quantities have been used in the production of babbitt and chemicals (approximately 3—4% each).

It is important to note the large potential for growth in the consumption of organotin compounds in the manufacture of plastics. This could result in a consumption of approximately 5000 tonnes of tin in plastics in the year 2000 (Heindl, 1970).

3.4 Uses of Tin

3.4.1 Tin and inorganic tin compounds

Tin is mainly used by industries producing tin-plate, solder, babbit, brasses and bronzes, pewter, printer's alloy (type metal), plastics, and tin chemicals. The largest single use of tin is in tin-plated steel, either by hot-dipping or by electroplating in a continuous process in which thin layers are deposited, and a different thickness can be applied on each side of the same sheet steel. In addition to its use in food and beverage packaging, tin-plate is used extensively in aerosol containers. Tin is also used in the transportation, machinery, electrical, plumbing, and heating trades and industries as solder, and in bearings and pipes.

Tin-lead solders contain from 2% tin for container-seaming to 63% for electrical connexions. In lead-free solder alloys, tin is alloyed with antimony, silver, zinc, or indium to obtain special properties such as higher strength or corrosion resistance. The largest quantities of solder are used in car radiators, air conditioners, heat exchangers, plumbing and sheet metal joining, container seaming, generating equipment, electronic equipment, and computers.

The copper-tin alloys are called bronzes. Phosphor-bronzes (5—10%) are the most important of the tin bronzes, the major applications being in marine and railway engineering.

Metals used for casting or lining bearing shells are classed as white bearing alloys, but are better known as babbitt. Babbitt

Table 1. Some applications of inorganic tin and its compounds

Compound	Application
tin metal	manufacture of tin plate, solders, bronzes, pewter, alloys, amalgams, chemicals
tin(IV) oxide	ceramic glaze opacifier, ceramic pigments
tin(IV) hydride	gas-plate tin on metal, ceramics
tin(II) acetate	catalyst
tin(II) chloride	electrotinning of steel strip, tin coating of sensitized paper, antisludge agent for oils, stabilizer of perfumes in soaps, additive for drilling muds, electroplating, catalyst in organic reactions
tin(II) fluoroborate	tin-plating baths
tin(II) fluoride	toothpaste and dental preparations
tin(II) 2-ethylhexoate	catalyst for polyurethane foam production and incurring silicone oil formulations
tin(II) oxalate	catalyst for coal hydrogenation, catalyst for acid-type esterification, transesterification or polyesterification
tin(II) oxide	manufacture of gold-tin and copper-tin ruby glass
tin(II) sulfate	immersion plating of steel wire, electrotinning strip, with copper sulfate for lacquer finishes
tin(II) tartrate	dyeing and printing of textiles
tin(IV) chloride	mordant in dyeing of silk, preparation of other inorganic and organic tins, manufacture of blueprint and other sensitized papers
sodium stannate	alkaline electroplating tin baths
sodium pentafluorostannite	dentifrice formulations

alloys are used in bearings in marine propulsion, rail and road transportation, compressors, motors, generators, and fans.

Type metals are lead-based alloys containing 1—25% antimony and 3—13% tin that are widely used in the printing trade.

Pewter, which contains 90—95% tin, 1—8% antimony, and 0.5—3% copper, is used in the production of a wide variety of household articles.

Special alloys using tin include dental amalgams, which are mainly silver-tin-mercury alloys; alpha-type titanium alloys that are used in aircraft, and zirconium alloys used in nuclear reactors.

Some of the manifold applications of inorganic tin compounds are listed in Table 1.

3.4.2 Organotin compounds

Worldwide production of organotin compounds, the fourth largest synthesis of organometallic compounds, was approximately 27 000 tonnes in 1976 (Midwest Research Institute, 1977). This use of tin, however, represents only about 0.8% of the total metallic tin consumed globally.

The annual growth rate is expected to be 10% per year for the next 10 years, so that, by 1986, worldwide production of organotin compounds would be approximately 63 000 tonnes (Midwest Research Institute, 1977). This projected market outlet depends on 2 features. Since 70% of the total production of organotin compounds is used to heat-stabilize poly(vinyl chloride) plastic products, growth in PVC production must continue to increase at a projected rate of 10—12% per year for the next 10 years. Second, organotin compounds are expensive in comparison with other heat stabilizers. Therefore, they must remain competitive, at least as far as performance is concerned.

Of the 4 categories of organotin compounds, dialkyltin derivatives are the most important commercially. They are used as heat and light stabilizers for PVC plastics to prevent degradation of the polymer during the melting and forming of the resin into its final products. In addition, dialkyltin derivatives have the unique property of protecting the plastic product from degradation during use. Other important commercial uses of dialkyltin derivatives are as catalysts in the production of polyurethane foam products and as vulcanizing agents for silicone rubbers. The trialkyl-tin derivatives, which account for approximately 10% of the total production of organotin compounds, are used in agriculture as non-systemic fungicides and acaricides. The tetraalkyltin derivatives are used as intermediates in the manufacture of other organotin compounds. The monalkyltin derivatives have a limited commercial use as heat stabilizers for PVC plastic films.

About 0.5—2.0% by weight of dialkyltin derivatives are required in the stabilization of rigid and flexible PVC plastics, including food-grade PVC for wrapping and containers. In particular, dibutyl- and dioctyltin compounds are important heat and light stabilizers for PVC. In the USA, the Food and Drug Administration (1971) has set specific levels for 2 dioctyltin derivatives that can be present in food, packaged in food-grade PVC wrapping and containers. These 2 chemicals are dioctyltin maleate and dioctyltin *S,S'*-bis(isooctylmercaptoacetate) (Piver, 1973).

Dialkyltin compounds such as dibutyltin diacetate or-dilaurate are used as catalysts in the production of polyurethane foams. Recently, dimethyltin compounds have also been found suitable for the stabilization of PVC plastics.

Organotin compounds have many other applications: (*a*) as antioxidants and anticracking agents to retard rubber deterioration and to stabilize chlorinated rubbers in chlorinated paints; (*b*) in transformers, capacitators, and cables as hydrochloric acid scavengers to prevent corrosion when chlorinated diphenyls are present (e.g., tetraphenyltin); (*c*) as anti-oxidants for textile oils; (*d*) as activators, stabilizers, and catalysts for polymers such as polyesters and silicone elastomers and as catalysts for the polymerization of olefins; (*e*) in the treatment of glass to increase crack resistance;

(*f*) in the treatment of fibreglass (with alkyl- and aryltin compounds) for adhesion to resins; and (*g*) as curing catalysts in the application of silicone to textiles and paper.

Organotin compounds have biocidal properties and are used; (*a*) as agricultural fungicides (triphenyltin acetate, triphenyltin hydroxide); (*b*) as general biocides (bis(tributyltin) oxide = TBTO) in paints; in the preservation of manila and sisal ropes, leather, textiles; to make fabrics mildew-resistant; for the protection of jute and jute bags; in wood preservatives, slimicides, and in the production of paper; (*c*) as bactericides and biostats such as disinfectants for use in hospitals and stables (tributyltin benzoate); (*d*) as helminthicides in poultry (dibutyltin dilaurate, tetraisobutyltin); (*e*) as nematocides (*p*-bromophenoxy triethyltin); (*f*) as herbicides (vinyltin compounds, e.g., trivinyltin chloride); (*g*) as rodent repellents (tributyltin chloride, triphenyltin chloride and acetate); (*h*) as molluscicides (triphenyl- and tributyltin compounds; (*i*) as ovicides (trialkyl- and triaryltin chlorides in combination with DDT or pyrethrins); (*j*) as antifoulants in ship paints and underwater coatings (triphenyl- and tributyltin compounds); and (*k*) as miticides (tricyclohexyltin hydroxide). Furthermore, triphenyltin compounds have been suggested as insect chemosterilants. More detailed information concerning new uses of organotin compounds can be found in the proceedings of a recent symposium (Zuckerman, 1976). A schematic presentation of the various uses of organotin compounds is given in Fig. 1.

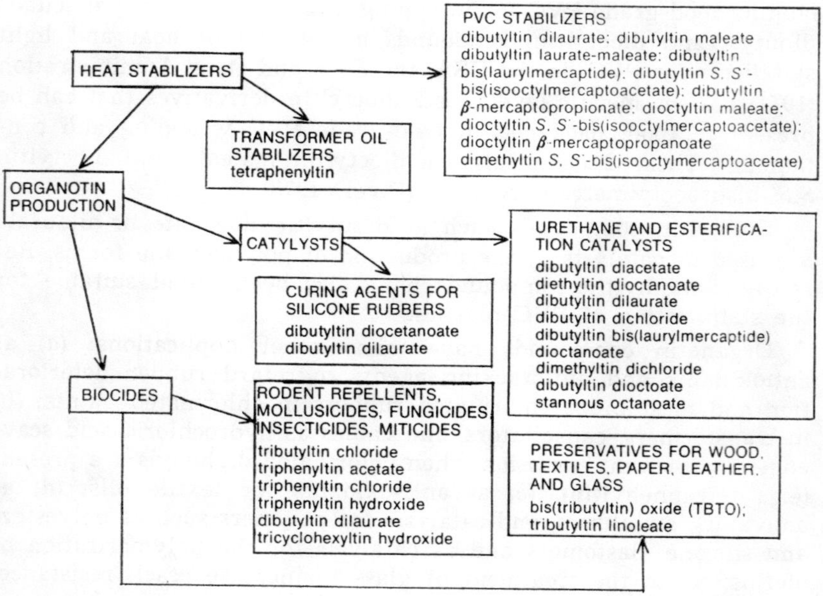

3.5 Tin-containing Wastes

Gases and fumes containing tin as well as sulfur dioxide and other contaminants, and both soluble and insoluble tin-containing waste materials such as water soluble salts, muds, and slags are produced during various smelting, refining, and detinning operations. Solid domestic and other wastes which may be dumped, incinerated, used for land fills or composting, contain much tin, and used cans, aerosol containers, and other miscellaneous tin-containing products in solid wastes account for 10—15% of tin-plated steel. The percentage of plastics in solid wastes is increasing yearly and includes PVC containing organotin compounds.

4. ENVIRONMENTAL TRANSPORT AND TRANSFORMATIONS

4.1 Transport and Bioconcentration

Tin concentrations in air are usually low except in the neighbourhood of some industrial sources (section 5.1). Similarly, it has not been consistently detected in all soils, plants, and waters (sections 5.2 and 5.3). This seems to support a geochemical classification placing tin in a group of elements with a low rate of migration in soils and waters (Perel'man, 1972; Bens et al., 1976). However, it is widely used and Wood et al. (1975) have included it among elements that are relatively accessible in the environment. A biological cycle for tin has been recently proposed by Ridley et al. (1977) (section 4.2).

There is a lack of information concerning the environmental transport of organometallic tin compounds. It appears, however, that the vapour pressures of some organotin compounds are high and that environmental mobility is possible although there are no data to confirm this deduction. Some organotin compounds seem to be well adsorbed in the soil and they have a low solubility in water (Heron & Sproul, 1958). A laboratory soil-leaching study (Barnes et al., 1973) indicated that triphenyltin is strongly attached to the soil. In natural waters, organotin compounds would be primarily adsorbed on the suspended particles and sediments (Schramel et al., 1973). More recent studies on the determination of tin and organotin compounds in natural waters have been reported by Braman & Tompkins (1979) and Hodge et al. (1979).

Little reliable information exists concerning the bioconcentration of tin and its derivatives. Schroeder et al. (1964) reported that the presence of tin in phosphate fertilizers originating from marine phosphate suggested that marine animals in the Pleistocene era absorbed tin from the sea. Although tin is present in seawater in concentrations of up to about 3 μg/litre (Mason, 1966; Vinogradov, 1953), there are few reports of its occurrence in marine algae, plankton, bacteria, flowering plants, protozoa, sponges, coelenterates, echinoderms, crustacea, and most fishes. However, Bowen (1966) noted tin levels of 0.2—20 mg/kg in certain marine organisms and accumulation of tin by the sponge *Terpios zeteki* was also reported by Bowen & Sutton (1951).

4.2 Environmental Chemistry of Tin

Because of the considerably higher toxicity of some organometallic tin compounds compared with inorganic forms of tin (section 7), the possibility of biomethylation of tin is obviously of considerable interest (Wood, 1974). A possible mechanism of such biomethylation has been proposed (Ridley et al., 1977; Wood et al., 1978). Recent laboratory studies have indicated that the methylation of tin by methylcobalamin[a] (CH_3—B_{12}) requires a one electron oxidation of Sn^{II} to a Sn^{III} radical, which can take place in the presence of Fe^{III} (Sn^{IV} would, of course, require a single electron reduction). The stannyl radical (Sn^{III}) can then react with CH_3—B_{12} (Co^{III}) to produce (under conditions of high chloride ion concentration) CH_3—$SnCl_3$ and reduced cobalamin, containing Co^{II} (Wood et al., 1978). A demonstration that methylation of tin takes place in a strain of *Pseudomonas* bacteria found in the Chesapeake Bay, USA, is the laboratory work with these bacteria carried out by Huey et al. (1974). However, the methylated tin species was not identified. Incubation of mercury(II) in the presence of tin(IV) resulted in enhanced formation of methylmercury, and the authors considered that methylmercury might have been transferred from the biologically methylated tin to the mercury(II) ion. Subsequently, Brinckman & Iverson (1975) proposed a "mercury-tin cross-over" system, which may have some real basis as indicated by Schramel et al. (1973), who found that mercury and tin accumulated together in some water plants in Bavarian rivers.

4.3 Degradation of Organometallic Tin Compounds

Organotin compounds may be degraded both chemically and biochemically. Hydrolytic decomposition occurs at rather extreme

[a] A form of vitamin B_{12} (cobalamin) in which the methyl carbon is directly bonded to the cobalt which is coordinated to the corrin ring system in the vitamin B_{12} structure.

pH values (≤ 1 or ≥ 13) unless there are other catalytic influences. Nevertheless, some authors consider that it could occur fairly rapidly in an aquatic environment, although the environmental pH is usually between 4 and 10 (Sheldon, 1975, Vizgirda 1972). This could be the case, if photochemical decomposition were also involved. Indeed, photochemical decomposition of triphenyltin acetate by ultraviolet irradiation has been demonstrated under laboratory conditions by Chapman & Price (1972) and Barnes et al. (1973). Most likely, under environmental conditions, the chemical and photochemical degradation of organotin compounds is combined with biochemical degradation, or degradation may be entirely biological. Some available information, mainly on triphenyltin, tricyclohexyltin, and tributyl tin compounds, is included in the following summary.

Brüggemann & Klimmer (1964) and Brüggemann et al. (1964a,b) reported that triphenyltin acetate on sugar beet leaves was rapidly broken down during the silage process. Within 5 weeks, triphenyltin acetate, originally present at a concentration of 2470 mg/kg of fresh leaves, had degraded completely. It was not established whether degradation was due to microbial action or resulted from the low pH (3.5—4.0) of the silage process. The authors also found that triphenyltin acetate was not broken down rapidly by microorganisms during its passage through the rumen and intestines of cattle. Similar results in feeding experiments with sheep were reported by Herok & Götte (1963).

Using thin-layer chromatography, Cenci & Cremonini (1969) found that triphenyltin acetate and hydroxide mixed into soil (80 mg/kg of soil) disappeared in 3—10 days, and 3 days, respectively, but the degradation products were not identified and the participation of microorganisms in the process was not established.

Akagi & Sakagami (1971) reported that, under their experimental conditions, ultraviolet irradiation of a solution of triphenyltin chloride yielded a mixture of triphenyl-, diphenyl-, and phenyltin as well as inorganic tin compounds, within 6 h. Similarly, irradiation of trialkyltin compounds resulted in a mixture of di- and monoalkyltin compounds and inorganic tin also within a period of 6 h. Total degradation of triphenyltin to inorganic tin required irradiation for more than 400 h. Chapman & Price (1972) investigated the degradation of an agricultural fungicide containing triphenyltin acetate, which had been found to disappear within a few days of spraying on crops, probably as a result of weathering and sunlight. They exposed the compound in thin layers on glass to ultraviolet light and found that the rate of degradation was higher at lower wavelengths and was independent of layer thickness (5—10 mμ). This was also true for the intermediate diphenyl- and phenyltin compounds and for the parent compound. When triphenyltin acetate las irradiated for 60 h (wavelength above 235 nm: intensity 120

W/m^2: thickness 10 μm), about 10% of the triphenyltin compound remained unchanged, about 30% occurred as diphenyltin, 15% as phenyltin compounds, and approximately 45% was degraded to inorganic tin in the form of hydroxides or hydrated oxides. Similar results were obtained by Barnes et al. (1973), who also studied breakdown at concentrations of 5—10 mg/kg using triphenyltin acetate in which the phenyl groups had been labelled with ^{14}C. Degradation was monitored by measuring the evolution of radioactive carbon dioxide; a half-time of about 140 days was determined. Since no $^{14}CO_2$ was evolved when the loam was heat-sterilized, it was concluded that the degradation process was due to microbial action.

The decomposition of triphenyltin chloride on sugar beet leaves, as reported by Freitag & Bock (1974b) gave the expected series of degradation products, the final degradation product being tin oxide. After 42 days, about 19% of triphenyltin chloride had undergone degradation.

In studies by Starnes (unpublished data)[a] and by Smith et al. (unpublished data)[b] quoted by Getzendaner & Corbin (1972), tricyclohexyltin hydroxide (the active ingredient of a commercial miticide) exposed on thin-layer chromatographic plates to a sun lamp yielded dicyclohexyltin oxide and cyclohexylstannoic acid with further degradation to inorganic tin. It has also been shown that tin residues on apples and pears treated with tricyclohexyltin hydroxide (commercial miticide) remained almost constant at a level of 0.1—0.2 mg/kg (Getzendaner & Corbin, 1972). This concentration was relatively independent of the total number of applications and, within a month of the final application, independent of the time between the last application and harvest.

Mazaev et al. (1976) reported the degradation of bis(tributyltin) oxide (TBTO), dibutyltin bis isooctylmercaptoacetate (BuIOMA), tributyltin methacrylate, diethyltin dioctanoate and dioctyltin bis (isobutylmaleate), and provided estimates for the half-times of these compounds for different aqueous media. In distilled water at 20 °C, values ranged from 1.14 days (BuIOMA) to 18.2 days (TBTO). From this report, it appears that dialkyltin compounds are more rapidly degraded than trialkyltin compounds.

Triorganotin compounds such as TBTO and tributytin fluoride used as antifouling agents in paints for ship bottoms and in underwater coatings appear to break down in a multistep process to give tin(IV) oxide (Barnes et al., 1973; Sheldon, 1975).

The mechanism of the biological dealkylation of organotin compounds is apparently more complicated than was considered by

---

[a] Starnes (1966) Unpublished report, The Dow Chemical Co. (ALS 66—648).

[b] Smith et al. (1970) Unpublished report, The Dow Chemical Co. (OL 30445 Feb. 1. 1970).

Blair (1975), who thought that the basic process was the oxidative cleavage of the carbon-tin bond (section 6.2.4).

Available information shows that the persistence of organotin compounds may vary considerably depending on the conditions and the type of compound. Under extreme laboratory conditions, the solvolysis may have a half-time ranging from about 1 min to about 100 days (Bassindale et al., 1971; Roberts & El Kaissi, 1968). Under environmental conditions, it is likely that the half-times for triphenyltin compounds are somewhere between a few days and 140 days (Freitag & Bock, 1974b). Diorganotin compounds probably have somewhat shorter half-times (Mazaev et al., 1976).

5. ENVIRONMENTAL CONCENTRATIONS AND EXPOSURES

5.1 Ambient Air

Tin is rarely detected in air, and, when detected, it is generally in low concentrations (0.01 $\mu g/m^3$), except in the proximity of some industrial sources. Tin emission concentrations of 10—640 $\mu g/m^3$ were reported from electric furnaces at certain plants in Japan, in 1972. At a distance of 700 metres, the atmospheric tin concentrations still ranged from 3.8 to 4.4 $\mu g/m^3$ (Environment Agency, Japan, 1971, unpublished data).[a]

Concentrations from 0.003 to 0.3 $\mu g/m^3$ were found in 60.6% of 754 samples tested from 22 cities in the USA. More than 50% of samples from 3 urban and 3 rural sites were below the detectable level. The highest concentration of tin (0.8 $\mu g/m^3$) was found in a sample from a Boston, USA, industrial site, which also contained the highest concentration of lead together with relatively high concentrations of zinc and cadmium (Tabor & Warren, 1958).

At stations of the US National Air Surveillance Networks during 1968 and 1969, concentrations ranged from below the minimum detectable level to 0.23 $\mu g/m^3$ in 1968 and 0.12 $\mu g/m^3$ in 1969. Both these concentrations were recorded in East Chicago (US Environmental Protection Agency, 1973).

Concentrations ranging from 0.04 to 0.09 $\mu g/m^3$, determined by emission spectroscopy, were reported for Cincinatti and St Louis in 1970 by the US National Air Surveillance Cascade Impactor Network (Lee et al., 1972). The size distributions of tin-containing particles found in this study are given in Table 2.

---

[a] Japanese Background Paper No. 2, prepared for the WHO meeting on the Effects on Health of Specific Air Pollutants from Industrial Emissions, Geneva, November 4—9, 1974.

Table 2. Size distribution of tin-containing particles in urban air [a]

	Cincinatti	St Louis	
	4th Quarter 1970	(A) 1st Quarter 1970	(B) 4th Quarter 1970
Average concentration ($\mu g/m^3$)	0.09	0.04	0.05
Average mass median diameter (μm)	0.93	1.40	1.53
% particles less than or equal to 1 μm	55	34	28
% particles less than or equal to 2 μm	86	68	65

[a] Adapted from Lee et al. (1972).

Peak concentrations of tin were found in particles ranging from 1 to 3 μm in diameter. Bogen (1973) found that ambient air levels of tin in the Heidelberg area of the Federal Republic of Germany between 30 April 1971 and 21 May 1971 ranged from 0.096 to 0.167 $\mu g/m^3$.

The combustion of coal, oil, and lignite results in the discharge of trace amounts of many elements into the atmosphere. Bertine & Goldberg (1971) pointed out that the principal sites of fossil fuel consumption are in the mid-latitudes of the Northern Hemisphere and suggested a need for linking fossil fuel consumption with the sedimentary cycles of trace metals from the atmosphere. They reported that tin was present in coal and oil at average concentrations of 2 and 0.01 mg/kg, respectively.

5.2 Soils and Plants

Tin has not been consistently found in all soils and plants; however, allowance must be made for the possibility that the concentrations present may have been below the limits of detection of the methods employed. Bowen (1966) reported levels of tin in soil ranging from about 2 to 200 mg/kg, the metal being strongly adsorbed by the humus.

Schroeder et al., (1964) observed that concentrations of tin in soils were localized and that it had not been detected in many areas. In a limited study of tin in vegetation and foods, he found that absorption by plants was erratic, even when the soil tin content was high. A tin concentration of 157 mg/kg dry weight was recorded in forest soil from southern Vermont, USA, but tin concentrations in sections of an 100-year-old elm indicated that exposure of this tree was not of recent origin.

Tin concentrations of 30—300 mg/kg have been reported in peat from Finland by Gordon (1952), and high concentrations of tin were found by Goldschmidt (1958) in forest litter and humus and also in coal. Tin was present in all but 7 of 43 samples of foliage from 13 species of ornamental plants and trees in New Jersey, USA (Hanna & Grant, 1962).

Lounamaa (1956), using a spectrographic method that had a limit of detection of 10 mg tin/kg of ash, detected tin only sporadically in higher plants in Finland. Lichens concentrated tin to exceptionally high levels, considering the small amounts present in rocks. For example, those growing on silicic rocks had tin concentrations of 72 ± 4.7 mg/kg ash (range: less than 10—100). Tin was undetected in 6 out of 16 mosses while the remainder had concentrations of 10—60 mg/kg of ash, concentrations being twice as high in those growing on silicic rock as in those growing on calcareous rock. Pasture herbage growing in Scotland was reported by Mitchell (1948) to contain tin concentrations of only 0.3—0.4 mg/kg dry weight. He also quoted tin concentrations in soils from north-east Scotland of up to 200 mg/kg. Warren (1964, unpublished data) reported a tin concentration of 75 mg/kg in soils in unmineralized areas of Devon and Cornwall in England, while soils from mineralized areas contained up to more than 1000 mg/kg.

5.3 Water and Marine Organisms

Tin has only been found occasionally in fresh water. It was found by Durum & Haffty (1961) in 3 out of 59 samples from 15 major North American rivers. No tin was detected by the National Water Quality Network (1960) in 119 analyses at 71 locations on 28 major rivers in the USA. Kleinkopf (1955, 1960) reported that Maine waters contained a mean tin concentration of 0.038 μg/litre with a maximum concentration of 2.50 μg/litre.

Tin has also only been found occasionally in municipal water supplies (Durfor & Becker, 1964). Analyses by the US Public Health Service showed concentrations of 0.0008—0.03 mg/litre in 32 out of 175 finished municipal waters (Taylor, 1964, personal communication).[a] Although it is known that acidic or alkaline corrosive waters can attack bronze fittings in pumping stations, presumably releasing some tin, the relationship between corrosion and the aqueous release of tin has not been established (Schroeder et al., 1964).

It has been reported by Vinogradov (1953) and Mason (1966) that tin is present in sea water in amounts of about 0.003 mg/litre; thus, its presence in marine organisms is to be expected. Tin levels of 0.2—20 mg/kg have been reported in marine animals by Bowen (1966) and accumulation by the sponge *Terpios zeteki* has been noted

---

[a] Taylor, F. B. (1964) Personal communication, cited by Shroeder, H. A., Balassa, J. J., & Tipton, I. H. (1964) Abnormal trace metals in man; Tin. *J. chron. Dis.*, 17: 494.

(Bowen & Sutton, 1951). Schroeder et al., (1964) reported that tin had not been detected in plankton, bacteria, flowering plants, protozoa, coelenterates, echinoderms, crustacea, or most fish. However, Vinogradov (1953) reported the presence of small amounts of tin in a marine worm, an oyster, and a dogfish.

Organotin compounds may enter water directly from antifouling coatings on ship bottoms or they may be used as molluscicides and added to vast areas of water for the control of snails, in which case the effective concentration has to be about 1 mg/litre (WHO, 1973). They may also be present indirectly as a result of industrial, agricultural, and other uses of organotin compounds.

Tin(IV) oxide, suggested to be the final breakdown product of some antifouling organotin agents (Sheldon, 1975), will eventually be deposited in bottom sediments because of its insolubility. However, no information is currently available concerning the rate and mechanism of this degradation.

5.4 Food

Tin has been reported to occur in trace amounts in most natural foods (de Groot et al., 1973). The tin content was determined by atomic absorption spectroscopy in 11 known wheats or wheat blends, in 20 flours prepared commercially from these wheats, and in 25 products specially prepared from the flours, as well as in 10 consumer products from 10 different cities in the USA. Only one half of the tin in bread could have been contributed by the flour. Tin concentrations in common hard, common soft, and durum wheat were 5.6 ± 0.6, 7.9 ± 0.9, and 6.8 ± 0.5 mg/kg respectively, while samples of flour and semolina from the above sources of wheat contained tin at 4.1 ± 0.4, 3.7 ± 0.7, and 6.0 ± 1.2 mg/kg, respectively. The tin concentrations in most products exhibited significant regional differences (Zook et al., 1970).

Schroeder et al. (1964) calculated that a diet composed largely of fresh meats, cereals, and vegetables would usually contain a tin concentration of less than 1 mg/kg and would supply about 1 mg of tin per day. In an area of southern Vermont, USA, where the soil contained considerable amounts of tin (about 160 and 33 mg/kg, respectively, in 2 gardens), 10 out of 18 samples of fresh vegetables contained tin levels of less than 1 mg/kg.

Hadžimicev (1971) measured the content of tin in foods from the vegetable belt around Sofia, Bulgaria. Tin concentrations in wheat, corn, beans, potatoes, tomatoes, cabbage, carrots, spinach, lettuce, onions, apples, and peaches ranged from 0.02 to 1.02 mg/kg.

Larger amounts of tin may be present in processed foods and drinks, because of corrosion and leaching of the metal from plain unlacquered cans or from tin foil used for packaging. The corrosion of tin-plate in food containers depends upon several factors including the type of food product, the time and temperature of storage, the acidity of the food and the quantity of air present in the headspace of the can (Calloway & McMullen, 1966; Monier-Williams, 1949). Oxidizing agents (eg., nitrates, ferric and cupric salts) and anthocyanin pigments, methylamine, sulfur dioxide, and other sulfur compounds accelerate corrosion, while tin salts in solution, sugars, and colloids such as gelatine retard it (Monier & Williams, 1949). The introduction of lacquered cans and the crimping of the tops minimizing direct contact of food with solder has resulted in a general reduction in corrosion and leaching.

A number of foods are unsuitable for packing in plain cans as they promote corrosion. Thus, tin has been found in canned asparagus in high concentrations varying from 120 to 550 mg/kg (Adam & Horner, 1937; Eyrich, 1972; Woidich & Pfannhäuser 1973). Tomato fruit can accumulate nitrate when grown under combined conditions of high temperature, high nitrogen fertilization levels, and low light intensity. Such tomatoes caused excessive detinning of internal plain can surfaces (Hoff & Wilcox, 1970). Iwamoto et al. (1968) also showed a distinct detinning effect on unlacquered cans of both naturally occurring and added nitrate in tomato juices; the authors suggested that the concentrations of nitrate-nitrogen in juices should not exceed 3 mg/kg. Tin contents exceeding 100 mg/kg have been found in other foods in unlacquered cans including fish, orange, grape, and mango juices, apricots, bananas, pineapple, and sugar syrup (Catala et al., 1971; Eyrich, 1972; Mahadevaiak et al., 1969; Stanculescu et al., 1972; Woidich & Pfannhäuser, 1973). In foodstuffs stored in all-lacquered cans, the tin content was mainly below 25 mg/kg (Catala, et al., 1971; Eyrich, 1972; Woidich & Pfannhäuser, 1973).

Several investigations have shown that high storage temperatures increase the transfer of tin into canned foods (Catala et al., 1971; Kimura et al., 1970; Zui, 1970). An increase of 2 mg/kg per month of storage was reported by Zui (1970) for an increase of 1 °C in temperature.

Nishijama et al. (1971) showed an increase in the tin content from 50—77 to 260—300 mg/kg in pineapples stored at 8 °C in unlacquered cans for 72 h after opening the can. The transfer of tin into the fruits was also increased by 0.5% aqueous citric or tartaric acid. The content of tin in foods in lacquered cans was low and remained low after opening. Klein et al. (1970) noted a similar effect of storage of fruit juices in opened plain cans. Values exceeding 200 μg/kg were found within 48 h of opening the cans.

A mean concentration of 0.0078 mg/litre was found in bottled (glass) cow's milk compared with 16 mg/litre in evaporated milk in unlacquered cans (diluted to the equivalent volume of cow's milk).

In some cases, concentrations of up to 110 mg/litre were found; however, only milk from unlacquered cans contained significant amounts of tin, while milk in lacquered cans contained less than 5 mg/litre (Hamilton et al., 1972). The highest concentration of tin found by Wodsak (1967) in canned condensed milk samples, less than 4 weeks old, was 40 mg/litre. The concentrations did not increase much during a further 5 months of storage but after 2 years of storage, concentrations up to 160 mg/litre were detected.

In addition to contamination by the containers, the presence of tin in food may be due to the use of tin as an additive. Tin(II) ions are used as additives in asparagus and peas packed in glass containers or in lacquered cans, and also in soda water. The stannous ion is added to prevent the migration of other heavy metals into the canned foods and to inhibit the oxidation of ascorbic acid. Scott & Stewart (1944) reported that *Clostridium botulinum* would not grow in beetroot and carrots in unlacquered cans whereas normal growth occurred in lacquered cans. For canned beetroot, the concentration of added tin required for the prevention of growth was 150 mg/kg and for carrots, 30—60 mg/kg.

High concentrations of tin have been reported in cheese packed in tin foil (Dyer & Taylor, 1931, Elten, 1929).

5.5 Organotin Residues

Tin occurring in food and beverages may also result from the use of organotin compounds as miticides in agriculture and as stabilizers in PVC materials. In several experiments in various parts of the USA, apples and pears from trees that had been sprayed 4 times with a miticide, tricyclohexyltin hydroxide, had maximum organotin residues of less than 2 mg/kg of whole fruit on the day of the final application. These concentrations were reduced to about half within 3—5 weeks and, after 4 weeks, the residues consisted mainly of tricyclohexyltin hydroxide with only small amounts of decomposition products; inorganic tin concentrations were almost invariably below 0.2 mg/kg. The mean concentration of inorganic tin on apples and pears after 1—5 applications of the miticide was 0.1 mg/kg (equivalent to 0.3 mg/kg, when calculated as tricyclohexyltin hydroxide); an inorganic tin residue of 0.3 mg/kg (equivalent to 1.0 mg/kg of tricyclohexyltin hydroxide) was found on pears by Getzendaner & Corbin 1972). In an appraisal of data on pesticide residues in food, the Joint Meeting of the FAO Working Party of Experts and the WHO Expert Group on Pesticide Residues (FAO/WHO, 1971) concluded that residues of tricyclohyxyltin hydroxide on apples and pears decline by 50% in about 3 weeks due to photodegradation (section 4.3). A 20—50% reduction can be achieved by washing and most of the residues can be removed by peeling the fruits, after which only 0.1 mg/kg may be expected in the fruit flesh. Data on tricyclohexyltin hydroxide on apples and pears after 1—4 treatments showed mean values ranging from 0.4 to 2.0 mg/kg. Similar concentrations on apples and pears have been reported from the Netherlands (WHO, 1975). It has also been reported that the concentrations of tricyclohexyltin hydroxide in treated products grown under glasshouse conditions including cucumbers, tomatoes, and bell peppers are unlikely to exceed 0.5 mg/kg (WHO, 1975).

The FAO/WHO joint meeting report (FAO/WHO, 1971) included data on triphenyltin hydroxide, acetate, and chloride residues in various foodstuffs such as potatoes, carrots, and sugar beets. Maximum concentrations in potatoes and carrots only rarely exceeded 0.1 mg/kg. Residues in food, fruit, and vegetables derived from direct contact with the compound could be considerably reduced by washing.

When cows were fed with sugar beet leaves containing triphenyltin acetate at 1 mg/kg a concentration of 0.004 mg/kg was found in the milk.

In some countries, regulations permit the presence of 2 PVC stabilizers in food, viz. dioctyltin *S,S'*-bis(isooctylmercaptoacetate) and dioctyltin maleate polymer, up to a concentration of total octyltin of 1 mg/kg (Department of National Health and Welfare, Canada 1975; Food and Drug Administration, USA, 1971). The migration of tin from PVC bottles into liquid foods contained in them was studied by Carr (1969). The increase in the tin content of various products during the storage period ranged from 0 to 0.07 mg/kg (Table 3).

Table 3. Tin concentrations in foodstuffs after storage for 2 months in PVC bottles at 30 °C

Food product	Tin content at beginning of experiment (mg/kg)	Tin content after ageing in PVC bottle (mg/kg)	Tin extracted from bottle (mg/kg)	Organotin stabilizer extracted from PVC bottle (mg/kg)
mineral water	0.076	0.088	0.01	0.063
tomato juice	0.03	0.03	0	0
peanut oil	0.05	0.06	0.01	0.063
vegetable oil	0.08	0.09	0.01	0.063
apple juice	0	0.02	0.02	0.126
cherry soda	0	0.07	0.07	0.443
beer	0	0.01	0.01	0.063
milk[a]	0.02	0.04	0.02	0.126
red wine	0	0	0	0
blended whisky	0.01	0.02	0.01	0.063

[a] Milk sample in PVC bottle aged for 2 weeks at 65 °C (Carr, 1969).

5.6 Working Environment

While most of the operations associated with the extraction and treatment of tin ore are wet processes, tin dust or fumes may escape during the bagging of concentrate in the ore rooms and during smelting operations (mixing plant and furnace tapping), as well as during the periodic cleaning of the bag filters used to remove particulate matter from smelter furnace flue gas before release into the atmosphere (ILO, 1972). Exposure to dust and fumes of tin oxide may cause stannosis.

Additional exposure can occur at the final stage of upgrading the cassiterite concentrate and during the roasting of sulfide ore. Tin(II) chloride constitutes a hazard, when the rough molten tin is separated from the rest of the charge during refining (ILO, 1972).

Other sources of industrial exposure include tinning of metals (Duckering, 1968); the preparation and use of tin alloys and solders; the use of tin(II) chloride as a reducing agent in calico printing; the use of hydrated stannic acid, sodium metastannate, hydrated tin(IV) chloride, and ammonium tin(IV) chloride as mordants in dyeing and in the weighting of silk; the use of tin(II) fluoroborate [$Sn(BF_4)_2$] in plating baths; and the production and use of organotin compounds.

5.7 Estimate of Effective Exposure of Man through Environmental Media

Food is the main source of exposure to tin for man. Estimates of human exposure from all environmental media are difficult to make because of the lack of reliable data concerning environmental concentrations (mainly in air and water), and because of variations in the amounts of different foods and beverages consumed, especially canned foods. Earlier estimates of the daily intake of tin by man have been difficult to reconcile with more recent estimates. Kehoe et al. (1940) reported that the mean daily intake of tin by a normal adult in the USA was 17 mg, while Tipton et al. (1966, 1969), in long-term balance studies, reported average daily intakes of between 1.5 and 8.8 mg in 4 subjects. Schroeder et al. (1964) found 3.6 mg in a day's institutional diet and considered that the major portion of the tin probably came from canned fruit juices and fruits in the diet. However, Schroeder and his colleagues pointed out that human intake may vary considerably. They calculated that a 10 MJ (2400 kcal) diet composed largely of fresh meats, grain products, and vegetables, which usually contain less than 1 mg/kg of tin, would supply 1 mg/day. However, a diet including a substantial proportion of canned vegetables and fish could supply as much as

Table 4. Daily tin balance for an adult in the USA[a]

Intake (mg)		Output (mg)	
food	4.0 (range, 1—40)	faeces (estimated)	3.98 (range, 1—40)
water	0.0 (range, 0—0.03)	urine	0.023 (range, 0—0.04)
air	0.003 (range, 0—0.007)		
total	4.003 (1—40)		4.003 (1—40)

[a] From: Schroeder et al. (1964).

Table 5. The concentration of tin in foods and total daily intake for an adult in the United Kingdom[a]

Item	Number of samples	Concentration of tin (mg/kg)[b]
cereals	3	$4.7 \pm 1.5 \times 10^{-2}$
meat	4	$2.1 \pm 0.5 \times 10^{-3}$
fats	4	$3.7 \pm 0.8 \times 10^{-2}$
fruits	5	0.5 ± 0.1
root vegetables	3	$2.2 \pm 1.0 \times 10^{-2}$
green vegetables	4	2.3 ± 10^{-2}
milk (cow's)	17	$7.8 \pm 1.2 \times 10^{-3}$ (mg/litre)
total daily untake		0.187 ± 0.042

[a] From Hamilton et al. (1972).
[b] For prepared diet.

38 mg/day. A typical daily tin balance for an adult in the USA, as estimated by Schroeder et al. (1964), is shown in Table 4.

Hamilton et al. (1972) reported a daily intake of tin from food and beverages by an adult in the United Kingdom of 0.187 ± 0.042 mg (Table 5). It seems likely that this diet was composed of fresh foods, which would account for the lower values obtained compared with the North American data.

6. METABOLISM

Although a substantial amount of literature exists on the absorption, distribution, excretion, and storage of tin, the results of much of the early research are questionable in the light of modern analytical techniques. For example, some studies depended on tissue matrix destruction processes with the risk of loss of tin salts by volatilization. In many cases, conditions were not controlled or defined making it difficult to know whether pure valence states of tin(II) or tin(IV) were being used, and whether the oxidation state of the element affected its biological fate. In most cases, the purity of the compound used was not indicated.

6.1 Inorganic Tin

6.1.1 Absorption

Evidence obtained from man and several animal species shows that ingested inorganic tin is poorly absorbed. Most studies indicate that less than 5% is absorbed from the gastrointestinal tract, although values as high as 20% have been reported (Furchner & Drake, 1976; Hiles, 1974; Kehoe et al., 1940; Kutzner & Brod, 1971; Monier-Williams, 1949; Schroeder et al., 1964).

The influence of oxidation state and anion complement on the rate of gastrointestinal absorption was studied by Hiles (1974) by administering tin to rats in the form of tin(II) citrate, fluoride, or pyrophosphate, and tin(IV) citrate or fluoride as a single oral dose at 20 mg/kg body weight. The absorption of tin(II) citrate and fluoride was calculated to be 2.8%, whereas only 0.6% of the dose of tin(IV) citrate and fluoride was absorbed; the citrate and fluoride were equally absorbed. Absorption of tin, when the anion was pyrophosphate, was significantly lower than when the anion was citrate or fluoride. The author considered that this was because pyrophosphate had a greater tendency to form insoluble complexes with tin than either fluoride or citrate.

6.1.2 Distribution

With both oral and parenteral administration of tin to animals, the highest concentrations were found in the kidney, liver, and bone (Durbin, 1960; Hiles, 1974; Moskalev, 1964), the principal site of deposition being the bone (Furchner & Drake, 1976; Hiles, 1974).

The amounts of tin remaining in the organs of rats 48 h after receiving a single oral dose of 20 mg of tin in the form of a radioactive compound are shown in Table 6; the distribution 48 h after a single intravenous administration of tin(II) or tin(IV) citrate is shown in Table 7 (Hiles, 1974).

Rats given a daily oral dose of 20 mg of tin per kg of body weight as tin(II) or tin(IV) fluoride did not show any detectable transplacental transfer at day 10 after conception, and at day 21, the amounts of tin found in rats given the tin(II) compound were so close to the limit of detection (= 0.1 μg of tin) that the significance of the findings remained doubtful. In rats receiving the same compounds at similar daily doses for 28 days, no significant amounts of tin could be detected in the brain, heart, testes, adrenals, reproductive tract, skeletal muscle, and spleen, while the levels in kidney and liver were approximately the same as after a single oral dose; however, the levels in bone were about 8 times higher than those found after a single oral dose. Tissue concentrations were higher

Table 6. ^{113}Sn distribution in rats after a single oral dose of solutions of various ^{113}Sn compounds[e]

Sample	Percent of dosed radioactivity				
	SnF_2	SnF_4	Sn(II) Citrate	Sn(IV) Citrate	$Sn_2P_2O_7$
Faeces and gastrointestinal wash	100.60±9.9 (10) [a]	98.40±3.6 (10)	98.00±2.1 (5)	95.40±10.3 (5)	100.0±2.8 (6)
Urine	1.02±0.58 (10)[b, d]	0.22±0.13 (10)[b]	0.90±0.23 (5)[c]	0.33±0.12 (5)[c]	0.41±0.19 (6)[d]
Kidneys	0.06±0.04 (10)[b, d]	0.01±0.01 (10)[b]	0.05±0.02 (5)[c]	0.01±0.00 (5)[c]	0.03±0.02 (6)[d]
Liver	0.06±0.04 (10)[b, d]	0.01±0.01 (10)[b]	0.07±0.02 (5)	< 0.004 (5)	0.02±0.02 (6)[d]
Pancreas	< 0.004 (5)	0.02±0.03 (5)	< 0.004 (5)	< 0.004 (5)	< 0.004 (5)
Recovery	101.74±9.4 (10)	98.66±3.6 (10)	99.02±2.3 (5)	95.74±10.5 (5)	100.46±2.9 (6)
	Percent of dosed ^{113}Sn/g of sample				
Bone (femur)	0.20±0.13 (10)[b, d]	0.04±0.02 (10)[b]	0.09±0.04 (5)[c]	0.03±0.02 (5)[c]	0.02±0.01 (6)[d]
Lymph nodes	< 0.004 (5)	0.08±0.12 (9)	< 0.004 (5)	< 0.004 (5)	0.01±0.01 (6)
Blood clot	0.01±0.01 (10)	< 0.004	0.01±0.00 (5)	< 0.004 (5)	< 0.004 (6)
Leg muscle	< 0.004 (5)	0.01±0.01 (5)	< 0.004 (5)	< 0.004 (5)	< 0.004 (6)

[a] % of dosed $^{113}Sn \pm SD$ (number of animals) where 1% = 40 μg. Tissues containing < 0.01% of the dose were not normally included in the table.
[b] Significant difference at $p < 0.05$ between SnF_2 and SnF_4.
[c] Significant difference at $p < 0.05$ between Sn(II) citrate and Sn(IV) citrate.
[d] Significant difference at $p < 0.05$ between SnF_2 and $Sn_2P_2O_7$.
[e] From: Hiles (1974).

Table 7. Fate of inorganic tin in rats. Tin distribution after a single intravenous dose of ^{113}Sn(II) or ^{113}Sn(IV) citrate[e]

Sample	Percent of dosed radioactivity			
	Sn(II) citrate		Sn(IV) citrate	
	Intact	Bile-duct cannulated	Intact	Bile-duct cannulated
Urine	35.3±4.6[a]	23.3±12.8	39.8±10.8	24.8±2.5
Faeces	12.1±1.8[b, d]	2.2±1.1[d]	3.1±1.4[b]	0.5±0.3
Bile	—	11.5±2.4[c]	—	0.1±0.1[c]
Injection site	3.3±1.8	6.2±4.7	4.1±1.5	3.9±3.3
Liver	2.0±0.2	5.1±1.6[c, d]	0.2±0.1[b]	0.2±0.1[c]
Kidney	5.9±0.9	8.3±1.3	5.3±3.4	4.2±2.5
Lungs	0.12±0.08	0.2±0.1	< 0.04	< 0.06
Spleen	0.4±0.0	—	< 0.04	—
Pancreas	0.2±0.1	—	< 0.04	—
Stomach	0.1±0.1[d]	0.2±0.0[d]	< 0.04	< 0.06
Small intestine	0.5±0.1[d]	0.4±0.4[d]	< 0.04	< 0.06
Large intestine	1.0±0.8[b, d]	0.4±0.3[d]	0.2±0.3[b]	< 0.06
Carcass	33.4±5.3	—	43.2±5.3	—
Femur (one)	1.6±0.2	1.9±0.1	3.2±0.6	2.0±0.1
	Percent of dosed ^{113}Sn/g			
Bone (femur)	4.5±0.6[b, d]	8.4±1.0[d]	8.8±1.8[b]	7.4±0.9
Blood clot	0.6±0.2[d]	0.9±10.3[d]	< 0.04	< 0.06
Leg muscle	0.06±0.03	—	< 0.04	—

[a] % of dosed $^{113}Sn \pm SD$ where 1% = 4 μg tin.
[b] Significant difference at $p < 0.05$ between intact Sn(II) and Sn(IV) animals.
[c] Significant difference at $p < 0.05$ between bile-duct cannulated Sn(II) and Sn(IV) animals.
[d] Significant difference at $p < 0.05$ between Sn(II) intact and bile-duct cannulated animals.
[e] From: Hiles (1974).

following administration of tin(II) fluoride than following tin(IV) fluoride administration (Hiles, 1974).

The results of distribution experiments performed by Hiles (1974) suggested that tin is unlikely to be rapidly oxidized or reduced during absorption and systemic transportation, although the actual oxidation state of the tin in the tissues could not be determined.

Following intravenous injection of ^{113}Sn(II) and ^{113}Sn(IV) citrates in rats, only small quantities of tin could be detected in lung tissues (Hiles, 1974). However, tin is found in the human adult lung and there is evidence indicating that the concentrations increase with age often reaching higher levels than those in other tissues (Schroeder et al., 1964). Thus, it seems probable that tin reaches the lung primarily through inhalation.

Levels of ^{113}Sn in the blood of rats, 2 days after oral or intravenous administration, were extremely low and were only detected in the erythrocytes (Hiles, 1974). Kehoe et al. (1940) found that 80% of the tin in the blood of man was in the cells.

6.1.2.1. *Distribution in human tissues and biological fluids*

Tin concentrations found by Hamilton et al. (1972/1973) in some tissues of healthy human adults are given in Table 8 together with data reported by Kehoe et al. (1940) and Schroeder et al. (1964). Comparatively high concentrations were found in lung, kidney, liver, bone, and also in lymph nodes. Fairly high levels were reported by Storoževa (1963) in the teeth of 24 persons, i.e., a range of 0.5—1.7 mg/kg of ash.

Schroeder et al. (1964) reported a highly variable distribution of tin in human tissues with significant differences related to age

Table 8. Tin content of human tissues, concentrations expressed as mg/kg wet weight.

Tissue	Hamilton et al. (1972/73)	Kehoe et al. (1940)	Schroeder et al. (1964)
	mean ± S.E.	mean	range (8 cities in USA)
blood	0.009 ± 0.002		
brain	0.06 ± 0.01	ND [a]	
kidney	0.2 ± 0.04	0.2	0.23—0.76
liver	0.4 ± 0.08		0.35—1.0
lung	0.8 ± 0.2	0.45	0.49—1.20
lymph node	1.5 ± 0.6		
muscle	0.07 ± 0.01	0.1	
bone	4.1 ± 0.6 [b]	0.5—0.8	

[a] ND = not detected.
[b] mg of element per kg of ash.

Table 9. Tin in human tissues by age, mean US values (mg/kg ash)[a]

Age group (year)	0	0—1	1—10	11—20	21—30	31—40	41—50	51—60	61—70	71—84
kidney										
concentration	0	57[b]	60[b]	33	20	34[b]	28[b]	22	34	32
occurrence	0/16	9/10	19/20	10/15	17/21	36/43	22/26	26/27	13/14	8/9
liver										
concentration	0	48[b]	61[b]	42	34	33[b]	25	35	38[b]	34[b]
occurrence	0/21	11/12	22/23	14/14	20/21	39/43	27/28	25/25	13/13	9/9
lung										
concentration	(18)	35	34	45[b]	27	31[b]	39[b]	53[b]	58	64[b]
occurrence	1/10	5/6	5/6	10/13	18/21	34/35	27/27	28/29	13/13	9/9
ileum										
concentration	0	101	98	80	174	116	97	53	172	140
occurrence	0/8	1/2	1/1	9/10	12/13	26/28	16/17	12/12	6/6	3/4
total occurrence (%)	1.8	86.7	94.0	82.7	88.2	79.8	93.9	97.8	97.7	93.5

Note: Concentrations of positive samples.
Occurrence refers to the number positive/total cases.
[a] From: Schroeder et al. (1964).
[b] Excluding values > 150.

Table 10. Geographical distribution of tin in kidney, liver, and lung of man (mg/kg ash)[a]

	No. of samples	% present	Mean	Range	Median
kidney					
United States of America	161	97	30	< 5—480	20
Switzerland	9	33	5	< 5—17	< 5
Africa	53	19	5	< 5—30	< 5
Middle East	43	42	8	< 5—55	< 5
Far East	57	40	11	< 5—110	< 5
liver					
United States of America	163	96	35	< 5—300	24
Switzerland	9	78	6	< 5—20	Tc
Africa	49	25	4	< 5—20	< 5
Middle East	44	55	11	< 5—140	Tc
Far East	54	41	10	< 5—74	< 5
lung					
United States of America	159	98	69	< 5—920	39
Switzerland	7	100	37	23—62	28
Africa	50	60	10	< 5—40	Tc
Middle East	45	91	33	< 5—200	17
Far East	57	93	65	< 5—1200	29

Note: Tc = trace, or barely detectable.
Mean and median values were obtained by assigning a value of one-half the least detectable amount to those cases where tin was not detected.
[a] From: Schroeder et al. (1964).

and geographical location. Variations in the tissue contents of tin among several geographical regions throughout the world were quite marked but were less noticeable between 8 cities of the USA (Table 9 and 10). All kidney, liver, and lung samples from Africa, Asia, and Europe had lower mean and median tin contents than samples from the USA (the difference being least in lungs). Tin in the tissue of Africans, where exposure to tin would be expected to be much lower than in industrialized North American and European areas, was, indeed, noticeably low. Tin was rarely detected in the tissues of stillborn infants, indicating that tin does not readily cross the placental barrier. The lungs accumulated tin with advancing age but other organs did not. Concentrations of tin in kidney, liver, lungs, and the ileum according to decade of life are shown in Table 9. Small amounts were found in the heart, prostate, uterus, and trachea. Avtandilov (1967) studied the trace element contents of normal and atherosclerotic aortas. He reported a mean tin concentration of 0.08 mg/kg wet weight in 20—39-year-old subjects with no difference between the two categories of aorta.

Using a spark source mass spectrometric method with a detection limit of 0.004 mg/kg wet weight, Hamilton et al. (1972/1973) reported a mean concentration of 0.06 mg/kg (± 0.01) in 10 whole brain samples from accidentally killed, presumably healthy adults. The concentration found in the frontal lobe was 0.03 (± 0.07) and that in the basal ganglia was 0.04 (± 0.02) mg/kg, measured in 2 samples. Tin was not detected in human brain by Kehoe et al. (1940) and was found in only a small percentage of subjects by Tipton (1960) and Schroeder et al. (1964).

6.1.3 Excretion

The major route of excretion of absorbed inorganic tin is the kidney although a small fraction is excreted into the bile (Hiles, 1974; Moskalev, 1964). The excretion of tin in the urine was studied by Perry & Perry (1959) in 24 human adults. A mean of 16.6 μg/litre of urine or 23.4 μg/day was reported. Kehoe et al. (1940) reported a mean concentration of 18 μg/litre of urine in subjects in the USA. There remains some doubt, however, as regards the accuracy of the analytical technique used since, in the same paper, the concentration in the urine of French males was reported to be zero.

In an experiment on rats, Hiles (1974) reported that after a single oral dose of tin(II) or tin(IV)citrate or fluoride at 20 mg/kg body weight (as the metal), about 50% of the absorbed tin was excreted within 48 h. Following intravenous administration of ^{113}Sn (2 mg of tin per kg body weight), 12% of the ^{113}Sn(II) and only 4% of the ^{113}Sn(IV) appeared in the faeces of rats indicating that the biliary route is probably more important in the elimination of tin(II) than of tin(IV) compounds. Most of the biliary excretion (94%) took place within 24 h.

6.1.4 Biological half-time

A half-time of 3—4 months was reported for ^{113}Sn in the skeleton of rats after intramuscular administration (Hamilton,

1948). However, Hiles (1974) found a half-time of only 34—40 days for both tin(II) and tin(IV) in the bone of rats following oral administration of tin(II) fluoride or tin(IV) fluoride for 28 days. The half-time of tin(II) for liver and kidney was reported to be 10—20 days. Furchner & Drake (1976) using tin(II) chloride administered intraperitoneally and intravenously in the mouse, rat, monkey, and dog, described the elimination from the body as a 4 component-process that was similar in all the species studied. The half-time for the longest component was over 3 months.

6.2 Organotin Compounds

In this document, consideration will be limited to compounds of the general type $RSnX_3$, R_2SnX_2, R_3SnX, and R_4Sn where R is a simple or complex anion. Generally, little information is available on the absorption, distribution, and excretion of these substances.

6.2.1 Absorption

Absorption from the intestinal tract of tin compounds with short alkyl chains varies depending on the compound. There are also considerable differences in absorption between species (Barnes & Stoner, 1959).

Ethyltin trichloride administered orally was poorly absorbed by rats, 92% of an oral dose of 25 mg/kg being eliminated in the faeces within 2 days. It was not excreted in the bile (Bridges et al., 1967). Triphenyltin acetate was almost completely eliminated in faeces within a few days, when administered orally to sheep (Herok & Götte, 1964) and cows (Brüggemann et al., 1964a). However, it was well absorbed, when given orally to guineapigs and mice (Stoner, 1966). Results obtained with rats were conflicting (Klimmer, 1963, 1964; Stoner, 1966). Tricyclohexyltin hydroxide was poorly absorbed in rats, only about 2% being excreted in the urine after a single 25 mg/kg oral dose. The remainder was recovered from the faeces and biliary excretion did not occur (FAO/WHO, 1971).

Trialkyltin compounds were well absorbed on contact with the skin, since the dermal LD_{50}s of various trimethyltin derivatives in mice were of the order of 50—100 mg/kg (Hall & Ludwig, 1972) and that of bis(tributyltin) oxide was below 200 mg/kg in rats and mice (Ascher & Nissim, 1964; Elsea & Paynter, 1958). When a 20% fat solution of various triethyltin derivatives was applied to the skin of rats and mice for 10 min, all the animals died within 20—30 min (Ignatjeva et al., 1968). In contrast, triphenyltin acetate did not penetrate unbroken skin readily (Stoner, 1966).

6.2.2 Distribution

Following intravenous administration to mice and rats, the highest concentrations of dibutyl- and diethyltin were found in liver and kidney tissue; unchanged compounds were excreted in the bile (Barnes & Magee, 1958). Rats were fed with 11 mg of triethyltin hydroxide over a period of 89 days. At the end of this time, only 0.7 mg of the compound could be recovered; 40% of this was in the blood, 28% in the liver, and 29% in the skeletal muscle. Smaller amounts of triethyltin were found in the kidney, brain, heart, and spleen (Cremer, 1957).

Species variation in the distribution of triethyltin has been reported. When triethyltin (26 μg) was added to rat blood *in vitro*, most of it (23 μg) was recovered from the erythrocytes and none was found in the plasma, whereas in rabbit blood, triethyltin was more equally distributed between erythrocytes (9 μg) and plasma (17 μg) (Cremer, 1957). This could explain why triethyltin persists in the blood of treated rats, whereas it quickly disappears from the blood of treated rabbits (Barnes & Stoner, 1959). Triethyltin was found in the brain of rats only 1—2 h after intraperitoneal injection of triethyltin hydroxide (Cremer, 1957).

The tissue distribution of triethyltin following intravenous administration of tetraethyltin resembled that following administration of triethyltin (Cremer, 1957, 1958). Thus, a large quantity of triethyltin was found in the liver, with smaller quantities in the kidney, brain, and whole blood in a rabbit, 2 h after administration.

After oral or intraperitoneal administration of ^{113}Sn-labelled triphenyltin to guineapigs, about 80% of the administered activity was excreted in 10 days. The concentration of radioactive tin found in the brain of rats and guineapigs after oral and intraperitoneal administrations decreased with a half-life of several days. The form of tin in the brain was not unequivocally identified, but it was not in the form of free stannic ions nor was it necessarily triphenyltin (Heath, 1967). Herok & Götte (1964) gave 3 sheep 10 mg of triphenyltin acetate orally, every day for 20 days. The compound was labelled with ^{113}Sn. The animals were killed at intervals ranging from 8 to 198 days after the last dose. The highest concentration of ^{113}Sn was found in the liver but the brain also contained measurable amounts. In rats, any absorbed triphenyltin chloride was rapidly distributed through the tissues of the body including the brain. Tin was eliminated relatively slowly and could be detected in the brain 38 days after a single dose (Heath, 1963).

Only trace quantities of tricyclohexyltin hydroxide were found in the tissues of rats and dogs fed up to 12 mg/kg bodyweight per day, in the diet, for periods ranging from 45 days to 2 years (FAO/WHO, 1971).

6.2.3 Excretion

When ethyltin trichloride was administered intraperitoneally to rats, it was excreted almost exclusively in the urine; biliary excretion was negligible. When diethyltin was administered intraperitoneally it was eliminated as diethyl- and ethyltin in both faeces and urine. Diethyltin was excreted in the bile (Bridges et al., 1967). Rats that had initially received a diet containing triethyltin and had retained approximately 0.7 mg triethyltin in their tissues were then given a normal diet; 12 days later, no triethyltin could be detected in the tissues (Cremer, 1957). The route of excretion was not known.

The elimination of triphenyltin from the body is slow (Stoner, 1966). Its persistence in guineapigs is suggested by the similarity between the amounts consumed by those dying on diets containing 25 and 50 mg/kg and the acute oral LD_{50}. In studies on guineapigs using ^{113}Sn-labelled triphenyltin, the concentration of triphenyltin that had entered the brain after a single oral dose did not decrease during the first 10 days. In rats, triphenyltin disappeared more rapidly, having a half-time of about 3 days in the brain (Heath, 1963, 1965). Triphenyltin acetate labelled with ^{113}Sn was slowly excreted in the urine of lactating sheep given oral doses of the compound at the rate of 10 mg/day for 20 days. The milk contained small amounts of tin (about 1 μg/litre) in both organic and inorganic forms (Herok & Götte, 1964).

The biological half-time of tricyclohexyltin hydroxide in rats was 5—40 days, when the compound had been included in the diet over a prolonged period. The brain was one of the tissues from which the compound was removed most slowly (FAO/WHO, 1971).

6.2.4 Biotransformation

In some early studies on the metabolic degradation of organotin compounds, it was suggested that destannylation (carbon-tin bond cleavage) was the major result of this biotransformation. For example, triethyl tin was detected in the tissues of rabbits and rats given tetraethyltin intravenously (Cremer, 1957, 1958). The liver was the organ most active in this conversion. Bridges et al. (1967) reported that dealkylation of diethyltin occurred in both the gut and tissues of the rat. Although Herok & Götte (1963) found that after administration of triphenyltin acetate labelled with ^{113}Sn small amounts of inorganic tin were present in the milk of sheep, Stoner (1966) considered that triphenyltin was not readily transformed in the body. Dicyclohexyltin oxide and trace amounts of cyclohexylstannoic acid were identified as metabolites in rats and dogs treated with tricyclohexyltin hydroxide (FAO/WHO, 1971).

Blair (1975) concluded that tricyclohexyltin hydroxide undergoes metabolism in animals by scission of cyclohexyl groups from the atom.

More recent studies using tributyltin acetate showed primary biological oxidation reactions in the hydroxylation of carbon-hydrogen bonds that are α, β, γ and δ to the tin atom (Fish et al., 1976). This type of reaction was already assumed by Casida et al. (1971), although no carbon-hydroxylated metabolites could be identified when triethyltin derivatives were biologically oxidized *in vitro* to monoethyl derivatives.

Fish et al. (1975) treated tributyltin acetate with rat liver microsomes (a mono-oxygenase enzyme system) in the presence of NADPH, and obtained α and β hydroxylated metabolites in yields of 24% and 50%, respectively. An *in vivo* study with [1-^{14}C]-tetrabutyltin showed the occurrence of similar reactions in mice (Kimmel et al., 1977). The α-hydroxyl metabolite is unstable and undergoes a cleavage reaction at pH = 7.4 to give 1-butanol and a dibutyltin derivative. A β-elimination reaction in acidic media transforms the β-hydroxyl derivatives rapidly into 1-butane and a dibutyl derivative (Fish et al., 1976).

Studies on the *in vitro* metabolism of tricyclohexyltin compounds indicated that carbon-hydroxylation of the cyclohexyl group was a major metabolic reaction (Fish et al., 1976; Kimmel et al., 1977). When [^{113}Sn] -triphenyltin acetate was subjected to *in vitro* biological oxidation, the phenyl group was not biotransformed, but in *in vivo* studies, rats were found to metabolize triphenyltin acetate into diphenyl- and phenyltin derivatives (Kimmel et al., 1977).

7. EFFECTS ON ANIMALS

7.1 Inorganic Tin Compounds

Although tin is present in small amounts in most animal and human tissues, it is uncertain whether it is an essential element for mammals. However, recent results indicate that tin is an essential nutrient for the growth of the rat. Both inorganic and organotin compounds at concentrations similar to those present in feeds were found to stimulate growth rate in rats maintained on purified amino acid diets (Table 11) (Schwarz, 1971; 1974; Schwarz et al., 1970). The authors concluded that tin, as an essential element, could have a function at the active site of some metal-dependent enzymes; however, this has still to be confirmed.

Compared with most organotin derivatives, inorganic tin and its salts are not highly toxic, mainly because of their poor absorption

Table 11. The effect of tin(IV) sulfate on the growth of rats in a trace-element-controlled environment [a]

Compound	Dose level (μg of tin/kg)	No. of animals	Average daily weight gain	Increase (%)	p-value
control	—	5[b]	1.10±0.05[c]	—	—
tin (IV) sulfate	500	8	1.37±0.10	2	.02
tin (IV) sulfate	1000	8	1.68±0.10	53	.001
tin (IV) sulfate	2000	8	1.75±0.10	59	.001

[a] From: Schwarz et al. (1970).
[b] 2 control rats died during the 26—29 days of the experiment.
[c] mean ± standard error.

and rapid tissue turnover (Barnes & Stoner, 1959; Cheftel, 1967; Hiles, 1974; National Academy of Sciences, Washington 1973). The systemic toxicity of some simple tin salts is difficult to assess because of the irritant properties of their solutions.

7.1.1 Effects on the Skin

The effects of 1% tin(II) chloride and 0.25% tin(II) fluoride solutions, applied to the abraded skin of rabbits, were examined by Stone & Willis (1968). Both compounds induced intraepidermal pustules with complete destruction of the epidermis but the stratum corneum remained intact. No injury occurred when the solutions were applied to intact skin.

7.1.2 Respiratory system effects

According to Robertson (1960), intratracheal administration of 50 mg of metallic tin dust to rats was well tolerated and no fibrosis was produced within one year of exposure.

Exposure of guineapigs by inhalation to tin(IV) chloride (3 mg/litre for 10 min, daily, for "several months") produced only transient irritation of the nose and eyes (Pedley, 1927).

7.1.3 Effects on the gastrointestinal system

Soluble tin salts are gastric irritants. This explains the signs of acute poisoning occasionally observed in man and in experimental animals following consumption of food containing high concentrations of tin (de Groot et al., 1973). However, the concentrations required to elicit an acute gastrointestinal reaction have not been determined reliably. Benoy et al. (1971) reported signs of illness in 3/10 cats after ingestion of 5 ml/kg body weight of fruit juice containing a tin concentration of 1370 mg/litre, and in 1/11 cats receiving the same dose of fruit juice containing tin at a concentration of 540 mg/litre. However, no adverse effects were recorded in cats receiving a similar amount of juice containing a tin concentration of about 500 mg/litre or in dogs given juice with a tin concentration of 1400 mg/litre.

7.1.4 Effects on the liver

A 3-fold increase in haem oxygenase (EC 1.14.99.3)[a] activity in the liver of rats was observed 16 h after a single subcutaneous injection of tin(II) chloride dihydrate ($SnCl_2 2H_2O$) at doses ranging from 5.6 to 56.4 mg/kg body weight. Cytochrome P-450 mediated drug metabolism and the content of cytochrome P-450 were reduced by one third. These effects on the liver increased by about 15—20%, when the compound was administered intraperitoneally (Kappas & Maines, 1976).

Tin chloride, oxalate, or sulfate added to the diet at a concentration of 10 g/kg for 4 and 13 weeks resulted in homogenous liver cell cytoplasm and hyperplasia of the bile duct in rats. The liver changes were more distinct after 13 weeks administration. Similar, although milder, hepatic alterations were seen after administration of tin chloride, oxalate, or orthophosphate at a concentration of 3 g/kg of diet[b] (de Groot et al., 1973).

Increased incidence of fatty degeneration in the liver was noted by Schrceder et al. (1968) in female but not in male rats, when the animals were given tin in the form of tin(II) chloride at a dose of 5 mg/litre in drinking water, from weaning until their natural death.

7.1.5 Effects on the kidney

Renal damage was produced in rats by single intravenous and intraperitoneal doses of sodium pentafluorostannite ($NaSn_2F_5$) and tin(II) chloride dihydrate ($SnCl_2 2H_2O$) (Conine et al., 1973, 1975; Yum et al., 1976). A single intraperitoneal injection of sodium pentafluorostannite at 35 mg/kg body weight or of tin(II) chloride dihydrate at 44.4 mg/kg produced extensive necrosis of epithelial cells mainly involving proximal tubules. The extent of the necrosis was thought to be related to the tin moiety of the compound because

[a] The numbers within parentheses following the names of enzymes are those assigned by the Enzyme Commission of the Joint IUPAC-IUB Commission on Biochemical Nomenclature.

[b] Tables for the approximate conversion of concentrations in diet to mg/kg body weight per day are given in Nelson (1954).

damage of the same magnitude could not be produced by administration of sodium fluoride (Yum et al., 1976).

Studies in which rats received single subcutanous injections of tin(II) chloride dihydrate in doses ranging from 5.6—56.4 mg/kg body weight revealed a 20 to 30-fold increase in the haem oxidation activity in the kidney, 16 h after administration. This effect was already noticeable at the lowest dose (5.6 mg/kg) and was found to be dose-related (Kappas & Maines, 1976).

Conine et al. (1976) administered daily oral doses of sodium pentafluorostannite for 15 or 30 days to rats at rates of 20, 100, and 175 mg/kg body weight. The highest dose caused degenerative changes in the proximal epithelium of the kidneys that were similar to those observed in chronic fluoride poisoning (Lindemann et al., 1959), and affected about 15—20% of the 30 rats that were killed. Most of the 8 animals in the 175 mg/kg group that died spontaneously during the experiment displayed necrosis of the proximal tubular epithelium similar to that previously reported to be caused by tin (Yum et al., 1976).

Exposure of rats to tin(II) chloride at a concentration of 5 mg of tin/litre of drinking water, for life, produced vacuolar changes in the renal tubules of animals of both sexes (Schroeder et al., 1968).

7.1.6 Effects on the blood-forming organs

When tin(II) (as chloride, orthophosphate, sulfate, oxalate, or tartrate) was given in the diet to Wistar rats of both sexes at concentrations of 3 and 10 g/kg for 4 weeks, food intake was reduced, growth was retarded and slight anaemia developed. The signs of anaemia included a reduction in haemoglobin levels of about 10% and corresponding reductions in haematocrit values, erythrocyte counts, and serum iron concentration (de Groot et al., 1973). Dietary supplements of iron had a markedly protective effect (de Groot, 1973; de Groot et al., 1973). The authors suggested that tin compounds might inhibit haemopoiesis, possibly by interfering with the intestinal absorption of iron. An alternative mechanism by which tin salts could cause mild gastrointestinal bleeding was not investigated.

A dose-related decrease in haemoglobin concentrations was noted in rats after daily oral treatment, for 15 days, with sodium pentafluorostannite at 100 and 175 mg/kg body weight. Administration of 20 mg/kg per day did not affect the haemoglobin level (Conine et al., 1976).

7.1.7 Central nervous system effects

High doses of inorganic tin compounds seem to affect the central nervous system producing such effects as ataxia, muscular weakness, and the depression of the central nervous system. These signs were observed by Conine et al., (1975) after oral administration to rats of sodium pentafluorostannite and tin(II) chloride dihydrate in single doses of the order of the median lethal dose (LD_{50}). According to de Groot et al. (1973), feeding rats with tin(II) chloride at a concentration of 10 g/kg diet for 13 weeks, resulted in a spongy state of the white matter of the brain. Ataxia, hind-leg paralysis, and death were also observed in rats by Mamontova (1940) following daily subcutaneous injections of 3 mg of tin(II) citrate for 5—6 months.

7.1.8 Effects on the reproductive system and the fetus

There is one report indicating that tin(II) chloride might have an effect on the reproductive system. De Groot et al. (1973) noted testicular degeneration in rats after prolonged feeding with this compound (10 g/kg of food, for 13 weeks).

Inorganic tin compounds have not been shown to be fetotoxic. Theuer et al. (1971) gave groups of pregnant rats sodium pentafluorostannite, sodium pentachlorostannite, and tin(II) fluoride corresponding to tin levels in the diet of 125, 250, and 500 mg/kg. The rats were killed on day 20 of gestation. No effects were seen in the fetuses; tin concentrations in the fetuses were about 1 mg/kg compared with approximately 0.65 mg/kg in control fetuses, indicating a rather low transplacental transfer.

7.1.9 Carcinogenicity and mutagenicity

Few reports are available concerning the carcinogenicity of inorganic tin compounds. Walters & Roe (1965) administered either sodium chlorostannate at 1 or 5 g/litre of drinking water or tin(II) oleate at 5 g//kg of diet to mice, for up to one year; the survivors were then killed. The incidence of lymphomas, hepatomas, or pulmonary adenomas did not increase with any of the regimens.

Roe et al. (1965) reported 3 malignant tumours in 30 August rats that survived for 1 year or more on a diet containing sodium chlorostannate at a concentration of 20 g/kg, whereas a control group of 33 rats did not exhibit any case of malignant tumours. One tumour was an adenocarcinoma of mammary origin, another a pleomorphic sarcoma in the uterus, and the third, an adenomacarcinoma in the jaw region. The difference was not statistically significant. No tumours were seen in another group of 27 rats surviving on a diet containing tin(II) 2-ethylhexoate at a concentration of 5—10 g/kg.

Administration of tin(II) chloride to rats and mice at 5 mg/litre in drinking water throughout their life-time did not produce any increase in the incidence of tumours compared with a control group consisting of an equal number of animals (Kanisawa & Schroeder, 1969).

Studies concerning the mutagenicity of inorganic tin compounds were not available to the Task Group.

7.1.10 Other effects

Growth retardation in rats has been reported after the administration of high doses of various tin compounds. Dietary levels of tin as tin(II) oxalate, orthophosphate, chloride, sulfate, and tartrate at 3 and 10 g/kg for 4 weeks caused inhibition of growth in rats and oedema and atrophy of the pancreas (de Groot et al., 1973). Daily administration, by gavage, of sodium pentafluorostannite at 100 and 175 mg/kg body weight, for 30 days, resulted in a dose-related retardation of growth in rats (Conine et al., 1976). Administration of tin(II) chloride at a concentration of 5 mg of tin per litre of drinking water reduced the life span of female rats, whereas the longevity of male rats was unaffected (Schroeder et al., 1968).

7.1.11 Effective doses and dose rates

7.1.11.1 *Lethal doses*

The median lethal doses (LD_{50}) for 2 inorganic tin compounds are given in Table 12. A marked difference between the oral and parenteral LD_{50} values can be explained by the low absorption of tin compounds.

Table 12. Median lethal doses (LD_{50}) of tin(II) chloride dihydrate ($SnCl_2 2H_2O$) and sodium pentafluorostannite ($NaSn_2F_5$) [a]

Route	Sodium pentafluorostannite			Tin(II) chloride dihydrate male rats (LD_{50}, mg/kg)
	male mice (LD_{50}, mg/kg)	male rats (LD_{50}, mg/kg)	female rats (LD_{50}, mg/kg)	
intravenous	18.9	12.9	12.9	29.3
intraperitoneal	80.9	75.4	65.0	258.4
oral (fasted)	— [b]	223.1	218.7	2274.6
(fed)	592.9	573.1	— [b]	3190.1

[a] From: Conine et al. (1975).
[b] no data.

Daily intravenous administration of 3.3—4.2 mg of sodium tin citrate per kg body weight provoked death in rabbits in 7—8 days, whereas at a dose of 2.5 mg/kg, death occured only after 15 or more days (Mamontova, 1940).

In studies on 6 rats, subcutaneous injection of 3 mg of tin citrate/animal, per day, killed all the animals in 5—6 months. Similar administration of the same compound in doses of 2 mg/kg body weight produced vomiting, diarrhoea, weight loss, ataxia, hind-leg paralysis, and death within 5 months (Mamontova, 1940).

7.1.11.2 *Minimum effective and no-observed-effect doses*

Some information on effective doses has been given in the sections describing the effects. However, the following information may also be of interest.

De Groot et al. (1973) reported a normal growth rate in Wistar rats that had received tin(IV) oxide, tin(II) sulfide, or tin(II) oleate at dietary levels up to 10 g/kg for 4 weeks; haematological data were also normal except for an increase in the haematocrit in male rats fed with the highest concentration of tin(II) sulfide. The weight and the gross and macroscopic appearance of the liver, kidney, heart, and spleen were also normal.

When a similar experiment was conducted with tin(II) chloride for 13 weeks, the no-observed-effect concentration in food appeared to be 1 g/kg, if the diet were supplemented with iron (de Groot et al., 1973). This is equivalent to a dose rate of about 20—30 mg tin/kg body weight per day, for 90 days.

No effect on growth was observed by Conine et al. (1976), when sodium pentafluorostannite was administered orally to rats at a dose rate of 20 mg/kg body weight per day for 30 days.

Oral administration of sodium tin citrate in dose rates ranging from 100—150 mg/kg body weight per day for 1 year did not produce any noticeable signs of poisoning in rats (Mamontova, 1940). Similarly rats fed a diet containing either sodium chlorostannate at 20 g/kg or tin(II) 2-ethyl hexoate at 5—10 g/kg for one year did not show any pathological changes in the gastrointestinal tract, kidneys, or liver (Roe et al., 1965).

Mice that received tin in the form of sodium chlorostannate at a concentration of 1 or 5 g/litre in drinking water or tin in the form of tin(II) oleate in the diet at a concentration of 5 g/kg during their lifetime did not show any adverse effects (Walters & Roe, 1965).

Mice and rats (Schroeder & Balassa, 1961; Schroeder et al., 1968) given tin(II) chloride in drinking water at a concentration of 5 mg of tin/litre grew normally throughout their life. The life span of mice of both sexes and of male rats was not affected but that of female rats was shorter and there was an increased incidence of fatty degeneration of the liver (section 7.1.4). Vacuolar changes in the renal tubules were apparent in rats of both sexes (section 7.1.5).

7.2 Organotin Compounds

Distinction should be made between the effects of di-, tri-, and tetrasubstituted organotin compounds. The principal toxicological difference is that some trisubstituted compounds have a specific effect on the central nervous system producing cerebral oedema (Barnes & Stoner, 1958; Torack et al., 1969), whereas disubstituted compounds do not produce this effect but are potent irritants that can induce an inflammatory reaction in the bile duct (Barnes & Magee, 1958). Toxicologically, the tetrasubstituted compounds resemble trisubstituted compounds, which are, generally, more toxic than the mono- and disubstituted derivatives.

7.2.1 Effects on the skin and eyes

Dermal application of dibutyltin dichloride at 10 mg/kg body weight per day for a period of 12 days, caused severe local damage and also bile duct injury in rats and mice. Guineapigs were more resistant, showing little reaction to daily applications of 120 mg/kg on 5 successive days (Barnes & Stoner, 1958).

An aqueous solution of bis(tributyltin) oxide applied to the shaved skin (30 × 60 mm) of rats at concentrations of 0.36—0.95 mg/kg body weight produced slight local irritation lasting 2—3 weeks (Pelikán & Černý, 1968b). At higher doses (1.40—185 mg/kg), marked inflammation developed into skin necrosis. Severe effects on the eyes were also observed with bis(tributyltin) oxide (Pelikán & Černý, 1969).

Triphenyltin hydroxide was reported not to irritate rabbit skin or sensitize the skin of guineapigs (Marks et al., 1969). However, it was found to be extremely irritating to the eyes, even after brief exposure, and could cause corneal opacity (Marks et al., 1969).

A solution of triphenyltin acetate in oil at a dose of 150 mg/kg body weight caused a skin reaction in rats (Klimmer, 1964). Doses of tricyclohexyltin hydroxide at 1.2—60 mg/kg body weight per day, applied to the skin of rabbits over a 3-week period, produced distinct reactions locally but did not result in any systemic ill effects (FAO/WHO, 1971, p. 527).[a]

---

[a] Unpublished reports on triphenyltin compounds and tricyclohexyltin hydroxide were abstracted in *1970 Evaluations of Some Pesticide Residues in Food* (FAO/WHO, 1971). The proprietary data were submitted to the (Footnote continued on p. 70)

7.2.2 Respiratory system effects

Pulmonary congestion and oedema were observed in rats after single intravenous administrations of solutions of diethyl-, dipropyl-, diisopropyl-, and dipentyltin dichloride (20 mg/kg body weight), in 0.05 ml of polyoxyethylene sorbitin monoleate, whereas in the case of dibutyltin dichloride, 10 mg/kg was enough to cause this effect (Barnes & Stoner, 1958). Petechial and ecchymotic haemorrhages in the trachea and larynx were observed in sheep after intrarumenal administration of tricyclohexyltin hydroxide at 500 mg/kg body weight. Congestion was present in the dorsal portion of all pulmonary lobes at a dose as low as 150 mg/kg (Johnson et al., 1975).

7.2.3 Effects on the gastrointestinal system

A single dose of butyltin trichloride, butyltin-*S,S′,S″*-tris(2-ethylhexylmercaptoacetate), butylstannoic acid, and butyl thiostannoic acid at 4000 mg/kg body weight, administered by stomach tube to mice, produced various degrees of submucosal, subserosal, and intraluminar gastrointestinal hemorrhage within 24 h (Pelikán & Černý, 1970b). Similar administration of dioctyltin-*S,S′*-bis(2-ethylhexylmercaptoacetate) produced dilatation of the stomach, which was filled with gas. The stomach walls appeared ischaemic whereas the intestinal walls were hyperaemic, and traces of blood were found in the contents of the small stomach. Similar, but more pronounced findings were recorded after administration of dioctyltin- bis-(butylmercaptoacetate), although no traces of blood were seen in the stomach. Treatment with dioctyltin bis(2-ethylhexylmercaptoacetate) resulted in large amounts of liquid in the stomach contents, and slightly hyperaemic stomach walls; other findings were similar to those produced by the other 2 dioctyltin compounds, dioctyltin bis(dodecylmercaptide) and octyltin tris(2-ethylhexylmercaptoacetate) (Pelikán & Černý, 1970a). Ingestion of dibutyltin dichloride at a dose of 50 mg/kg body weight per day for one week, caused a temporary dilatation of the stomach in rats due to accumulation of fluid, and diarrhoea (Barnes & Stoner, 1958). Intrarumenal administration of tricyclohexyltin hydroxide at 150 mg/kg body weight resulted in fluid diarrhoea in sheep which persisted up to death. Necropsy findings included mild enteritis and colitis characterized by severe hyperaemia and oedema (Johnson et al., 1975). Gastroenteritis occurred in rats receiving an oral dose of tricyclohexyltin

---

WHO by the manufacturers. Page numbers given in the text refer to the FAO/WHO publication, which also contains information with reference to authors and manufacturers.

hydroxide of 25 mg/kg body weight per day for 19 days (FAO/WHO, 1971, pp. 527—528).

7.2.4 Effects on the liver and bile duct

Steatosis of hepatocytes and enlargement of the liver were seen within 24 h of administration to mice through a stomach tube, of a single dose of butylstannoic acid, butyltin trichloride, butyltin tris(2-ethylhexylmercaptoacetate), or butylthiostannoic acid, each at a dose of 4000 mg/kg body weight (Pelikán & Černý, 1970b). In rats, a single oral dose of dibutyltin dichloride at 50 mg/kg body weight produced congestion and inflammation especially in the lower part of the bile duct (Barnes & Magee, 1958; Gaunt et al., 1968). When this dose of dibutyltin dichloride was given on 3 successive days, the lesion was more severe involving also the proximal part of the bile duct and the portal blood vessels. Necrotic areas were seen in the liver. In some cases, bile escaped from the injured duct into pancreas and peritoneum. The authors considered that death occurring 5 days after the 3 successive doses was the result of a general toxic effect of this compound, whereas death occurring later was secondary to bile duct and liver damage. Changes in the pancreas were inflammatory and confined to areas surrounding the bile duct. In surviving rats, examined 6—12 months after receiving 3 daily doses of 50 mg/kg, the bile duct was shorter and thicker than normal with fibrosis in the wall. A similar reaction in the bile duct was seen after a single intravenous administration of 5 mg/kg body weight, and after a dermal application of 10 mg/kg body weight. The effects on mice of oral doses of dibutyltin dichloride at 20—50 mg/kg body weight were similar to those seen in rats, although liver damage appeared to be more widespread. Repeated doses of this compound at 20—50 mg/kg body weight killed rabbits, but did not cause bile duct or liver injury; guineapigs, however, tolerated such doses without any signs of adverse effects (Barnes & Magee, 1958). It has been suggested that bile duct injury occurs only in species in which the bile duct and the pancreatic duct have a common course (Kimbrough, 1976).

Daily doses of dibutyltin dichloride at 0.1 and 1.0 mg/kg body weight, for 6 months, caused intoxication in rabbits with dystrophic changes in the liver (Mazaev & Korolev, 1969). Dioctyltin-*S,S′*-bis(2-ethylhexylmercaptoacetate), diootyltin-*S,S′*-bis(butylmercaptoacetate), and diootyltin bis(dodecylmercaptide)at an oral dose of 4000 mg/kg body weight produced steatosis of the hepatocytes in mice (Pelikán & Černý, 1970a). Nikonorow et al. (1973) reported a significant increase in the mean liver weight of rats after oral administration of dioctyltin-*S,S′*-bis(isoctylmercaptoacetate) at 20 mg/kg body weight per day, for 3 months.

A single oral dose of tributyltin acetate, benzoate, chloride, laurate, or oleate at 500 mg/kg body weight produced fatty infiltration of the liver in mice (Pelikán & Černý, 1968a) and triphenyltin acetate at a dietary level of 10 mg/kg produced fatty degeneration of the liver in guineapigs, when administered over a 2- year period (FAO/WHO, 1971, p. 341). Cholangitis was reported in rats after administration of tricyclohexyltin hydroxide at a dietary level of 400 mg/kg for 90 days (Shirasu, 1970). Similarly, both intra- and extrahepatic cholangitis was noted in rats that had received tricyclohexyltin hydroxide at 25 mg/kg body weight per day, for 19 days (FAO/WHO, 1971, pp. 527—528).

7.2.5 Effects on the kidney

Some degree of fatty degeneration of the renal cortical tubular epithelium was seen on histological examination of mice that had received a single oral dose (4000 mg/kg body weight) of butyltin tin-*S,S′,S″*-tris(2-ethylhexylmercaptoacetate), butylstannoic acid, or butylthiostannoic acid, but not in animals given butyltin trichloride. The structure of the renal tissue was unchanged. Macroscopically, all compounds produced slight hyperaemia in the kidneys (Pelikán & Černý, 1970b).

Dioctyltin-*S,S′*-bis(2-ethylhexylmercaptoacetate) and dioctyltin-*S,S′*-bis(butylmercaptoacetate) administered at the same rate and by the same route produced a similar slight fatty degeneration of the renal cortical tubular epithelium in mice (Pelikán & Černý, 1970a). Feeding rats with an organotin stabilizer, dioctyltin-*S,S′*-bis(isooctyl mercaptoacetate) at a dietary level of 200 mg/kg for 12 months produced an increase in kidney weight in female rats only (Nikonorow et al., 1973). Mazaev & Korolev (1969) also reported dystrophic changes in the kidneys in rats treated with dibutyltin dichloride at 0.1 and 1.0 mg/kg body weight.

Haemorrhages were reported in the kidneys of mice after single administrations (gavage) of tributyltin benzoate, chloride, laurate, and oleate at 500 mg/kg body weight. Hyperaemia of the kidney was seen in all groups, whereas tubular cells containing lipids were observed only in mice that had received laurate or oleate. Renal changes were not reported in mice that were similarly dosed with tributyltin acetate (Pelikán & Černý, 1968a). Congestion of the glomeruli and mild toxic nephrosis were observed in rats receiving a dietary level of tricyclohexyltin hydroxide of 400 mg/kg for 3 months (Shirasu, 1970). Similarly, toxic nephrosis was reported in rats after administration of tricyclohexyltin hydroxide at a daily dose of 25 mg/kg body weight for 19 days (FAO/WHO, 1971, pp. 527—528).

7.2.6 Effects on the lymphatic tissues and immunological effects

Small necrotic areas were observed in the germinal centres of the splenic follicles of mice after a single administration, through a stomach tube, of 4000 mg/kg body weight of butyltin-*S,S′,S″*-tris-(2-ethylhexylmercaptoacetate), butylthiostannoic acid, and butylstannoic acid, (Pelikán & Černý, 1970b); dioctyltin-*S,S′*-bis(2-ethylhexylmercaptoacetate), and dioctyltin-*S,S′*-bis(butylmercaptoacetate) (Pelikán & Černý, 1970a).

Seinen & Willems (1976) reported that the main action of dioctyltin dichloride, given to rats at dietary levels of 50 and 150 mg/kg, was on the thymus, the weight of which was significantly reduced during the experiment. In subsequent experiments, feeding rats with dioctyltin dichloride or dibutyltin dichloride at dietary levels of 50, or 150 mg/kg for 6 weeks, resulted in a dose-dependent reduction in the weight of the thymus and thymus-dependent peripheral lymphoid organs (Seinen et al., 1977a). The delayed type hypersensitivity was decreased and the allograft rejection was delayed when the same compounds were given in the same way to rats. Furthermore, it was reported that both these compounds decreased the survival rate of rat and human thymocytes in *in vitro* studies (Seinen et al., 1977b). The effects of diethyltin dichloride and dipropyltin dichloride on rat lymphoid organs were similar but less pronounced. However, other dialkyltin compounds such as dimethyltin dichloride, didodecyltin dibromide, and dioctadecyltin dibromide, as well as octyltin trichloride, trioctyltin chloride, and tetraoctyltin did not cause atrophy of lymphoid tissues (Seinen et al., 1977a). The rat proved more sensitive to the action of dioctyltin dichloride and dibutyltin dichloride on lymphoid organs than the mouse and the guineapig in which no atrophy of the lymphoid organs was produced. (Seinen et al., 1977a).

Studies on triphenyltin acetate have been reviewed by the FAO/WHO (1971, pp. 327—366). In guineapigs, a dietary level of 15 mg/kg for 77 days produced a decrease in plasma cells in the spleen and in the mesenteric, cervical, axillary, and popliteal lymph nodes. After feeding the same dose to female guineapigs for 104 days, a reduction in immune response to tetanus toxoid stimulation was observed. In another experiment, a decrease in the number of lymphocytes and leukocytes accompanied by histological changes in the lymphatic tissues, such as atrophy of the white pulp of the spleen, was seen in guineapigs at dietary levels of 5 mg/kg in females and 10 mg/kg in males administered, for 12 weeks. A decrease in the number of leukocytes was also reported in dogs after 8 weeks on a dietary level of triphenyltin hydroxide of 25 mg/kg and 8 weeks on 50 mg/kg.

7.2.7 Haematological effects

Decreases in the numbers of erythrocytes and reticulocytes were reported in rats after daily administration of an oral dose of dibutyltin sulfide at 0.1 mg/kg body weight for 6 months (Mazaev & Šlepning, 1973). A mild anaemia was noted in rats that had received dibutyltin dichloride for 3 months at a dietary level of 80 mg/kg, but a level of 40 mg/kg did not produce this effect (Gaunt et al., 1968).

In vitro studies of the haemolytic activity of trialkyltin compounds indicated that the most haemolytic organotin compounds were the alkyltin derivatives with alkyl groups of 3—6 carbon atoms (Byington et al., 1974). Decreases in the percentage haemoglobin and in the number of erythrocytes were reported in guinea-pigs treated with triphenyltin acetate at a dietary level of 10 mg/kg for 4 months (FAO/WHO, 1971, p. 337). A reduction in haemoglobin and leukocytes was also seen in rats after 12 weeks on a dietary level of triphenyltin hydroxide of 50 mg/kg (FAO/WHO, 1971, pp. 337—338).

7.2.8 Central nervous system effects

Intoxication in animals by lower trialkyltin compounds is manifested as generalized weakness progressing to paralysis, sometimes accompanied by generalized tremor. In rabbits, typical reactions to a lethal dose (5 mg/kg body weight) of triethyltin sulfate administered intravenously are prostration, flaccid paralysis, and encephalopathy (Stoner et al., 1955). These responses resemble those observed in human subjects, accidentally poisoned with triethyltin compounds. Daily intraperitoneal injections of triethyltin sulfate at 5 mg/kg body weight in saline solution resulted in death in rats within 3 days. Diffuse haemorrhagic encephalopathy was found in the brain (Suzuki, 1971). Rats maintained on a dietary level of triethyltin hydroxide of 20 mg/kg displayed weakness in the hind legs after one week. The weakness reached a maximum at 3—4 weeks, when about half of the rats died, but the signs disappeared in one week, when normal diet was restored. Some of the rats showed signs of becoming resistant to the substance. At a dietary level of 40 mg/kg, the recovery process was unaltered but, at 80 mg/kg, the rats developed generalized muscular tremors that resembled those seen in acute trimethyltin poisoning (Stoner et al., 1955). This course of events, including the development of resistance to triethyltin hydroxide at a dietary level of 20 mg/kg was corroborated by Magee et al. (1957), who was also able to produce a specific lesion of the central nervous system (CNS) by feeding rats with a dietary level of triethyltin hydroxide of 20 mg/kg. The lesion consisted of an oedema extending throughout the white matter. It was microscopically visible after 3 days of exposure and progressed to a maximum at about 2 weeks. After 4 months of normal diet, the changes in the brain and spinal cord disappeared. Electronmicroscopic studies of this lesion in rabbits disclosed that the myelin sheaths split to form clefts and vacuoles in which the fluid responsible for the oedema accumulated (Aleu et al., 1963). The triethyltin-induced lesion of the CNS is different from those produced by alkyl derivatives of lead, antimony, bismuth, and mercury, which cause damage to the nerve cells, but appears to be similar to that produced in rats by exposure to hexachlorophene (Kimbrough & Gaines, 1971). When 5 mg of triethyltin sulfate per litre of drinking water was administered to newborn rats and their mothers for 4 months, all young rats were asymptomatic, whereas the mothers showed paralysis of posterior limbs, severe cerebral oedema, and status spongiosus of the white matter (Suzuki, 1971). Hedges & Zaren (1969) described papilloedema in monkeys following a single intraperitoneal dose of triethyltin acetate of 0.1 mg/kg, body weight administered as a 5% solution. Oedema of the white matter of the brain extended through the chiasma and orbital optic nerve to the retrolaminar area, but the papilloedema was not due to extension of the intramyelinic tin oedema of the nerves through the lamina. Papilloedema does not develop in the cat when similarly treated, in spite of profound swelling of the intracerebral white matter, chiasma, and the orbital optic nerve as far as the retrolaminar portion.

Tricyclohexyltin hydroxide appears to be a depressant for the central nervous system, although it is not as potent as triethyltin acetate and cerebral oedema does not occur (FAO/WHO, 1971, p. 526). Johnson et al. (1975) studied the acute toxicity of tricyclohexyltin hydroxide in sheep by intrarumenal injection of doses ranging from 15 to 750 mg/kg body weight. Central nervous depression was observed at 50 mg/kg body weight.

Because of the enzymatic conversion of tetraalkyltin compounds by hepatic microsomal enzymes (section 6.2.4), these compounds act in a similar way to trialkyltin compounds. The toxicity of tetraalkyltin compounds in mice and dogs was studied by Caujolle and his colleagues (1954). Tetraethyltin was the most active, tetramethyltin slightly less toxic, and for the higher members of the series, the toxicity decreased with increasing molecular weight. The toxic effects of tetramethyltin differed from those of the other members of the series and the dominant findings were tremors and hyperexcitability. The major effects of the other members of the tetraalkyltin series were muscular weakness and paralysis followed by respiratory failure. Development of the signs of poisoning were noticeably slow following administration of tetraalkyltin compounds and death was often delayed. Intravenous administration of tetraethyltin at 25 mg/kg produced a slight increase in the respiratory rate and vasodilatation as immediate effects, but after 1.5—2 h, prostration with muscular weakness was noted. The later effects resembled those seen after administration of triethyltin; the mode of death was also similar (Stoner et al., 1955).

7.2.9. Effects on reproduction and the fetus

Few studies of the effects of disubstituted organotin compounds on reproduction have been reported. Nikonorow et al. (1973) reported toxic effects of dioctyltin-*S,S'*-bis(isooctylmercaptoacetate) on the embryo and fetus at daily oral doses of 20 and 40 mg/kg body weight, administered for about 3 months. A distinct increase in the incidence of fetal deaths was observed. Teratogenic effects were not found.

Investigations on triphenyltin hydroxide have yielded variable results. In 2 investigations on rats, oral dose rates of 20 mg/kg per day for less than 4 weeks produced histologically evident abnormalities in the testes and ovaries. Diminution in the size of the testes and less advanced maturation of the germinal epithelium was seen in rats receiving triphenyltin hydroxide at a rate of 5 mg/kg for 90 days, although a 3-generation reproduction study at this dietary level failed to demonstrate any adverse effect on the reproductive indices (FAO/WHO, 1971, pp. 332—333). In another experiment (Gaines & Kimbrough, 1968), dietary levels of up to 200 mg/kg over a 276-day period caused a reversible reduction in fertility in male rats which was considered to be due to a pronounced decrease in food intake, that was later reversed. No gross abnormalities were seen in the offspring of these males when mated with untreated females. In 3 later experiments (FAO/WHO, 1971, p. 333) in which weanling animals were fed the compound in their diets at doses of up to 25 mg/kg, the previously reported change in testicular development could not be confirmed.

In a 3-generation study in which rats produced 2 litters per generation, animals continuously received a dietary level of tricyclohexyltin hydroxide of up to 100 mg/kg (corresponding to 4—6 mg test compound/kg body weight per day for an adult rat). No evidence of any adverse effects on reproduction could be found and examination of the fetuses did not reveal any indication of teratogenic effects (FAO/WHO, 1971, p. 524). No evidence of teratogenic effects from tricyclohexyltin hydroxide was obtained in a study in which pregnant rabbits received up to 3 mg/kg body weight per day from the 8th to 16th day of gestation (FAO/WHO, 1971, p. 524). In a study in which diets containing tricyclohexyltin hydroxide were fed to Japanese quail, no evidence of ill effects on egg fertility or hatchability was seen at a dietary level of 1 mg/kg.

Effects observed at a dietary level of 10 mg/kg were of doubtful significance but a dietary level of 100 mg/kg had definite effects on fertility, egg production, hatchability, and embryonic mortality (FAO/WHO, 1971, pp. 523—524).

7.2.10 Carcinogenicity

In an 18-month study, mice were given an oral dose (stomach tube) of 0.46 mg/kg body weight per day of triphenyltin acetate between the ages of 7 and 28 days and thereafter a dietary level of 1206 mg/kg. There was no statistically significant increase in tumours compared with a control group (Innes et al., 1969). In a 2-year study on rats receiving tricyclohexyltin hydroxide at concentrations up to 12 mg/kg body weight (section 7.2.3), the pattern of tumour incidence throughout both the control and test groups appeared to be random and did not suggest a dose-response relationship (FAO/WHO, 1971, p. 527). Reports of experiments specifically designed to investigate the carcinogenicity of other compounds were not available to the Group.

7.2.11 Effects on chromosomes

Mazaev & Šlepnina (1973) reported an increased percentage of chromosomal aberrations in bone marrow cells and also an increased mitotic index in cells of the small intestine mucosa of rats that had received dibutyltin sulfide at 0.1 mg/kg body weight per day administered by intubation over a period of 6 months. Doses of less than 0.1 mg/kg body weight per day did not cause poisoning and did not have any effect on the chromosomes or somatic cells.

7.2.12 Other effects

A reduction in weight gain, mainly related to reduced food intake, has been recorded in various species given organotin compounds. In rats, a dietary level of dibutyltin dichloride of 80 mg/kg administered for 90 days caused growth retardation and decreased food intake (Gaunt et al., 1968). Triphenyltin acetate given to guineapigs at dietary levels exceeding 5 mg/kg for 2 years, caused a dose-related inhibition of growth rate (FAO/WHO, 1971, p. 341). A dietary level of tricyclohexyltin hydroxide of 25 mg/kg for 90 days, caused a slight reduction in weight gain in female rats (Shirasu, 1970), and dogs receiving the same compound in their diet at a level of 12 mg/kg for 6 months also lost weight; some animals that totally refused the food died from starvation (FAO/WHO, 1971, pp. 526—527).

Severe cardiovascular changes were demonstrated electrocardiographically in sheep after intrarumenal administration of tricyclohexyltin hydroxide at doses greater than 150 mg/kg body weight (Johnson et al., 1975).

7.2.13 Mechanisms of action

There have feen relatively few studies of the biochemical effects of dialkyltin compounds, which do not cause cerebral oedema. In acute toxicity experiments on the rat, diethyltin compounds did not affect the sodium and potassium contents of the central nervous system but caused a decrease in its water content (Magee et al., 1957).

The action of a homologous series of disubstituted organotin-compounds from dimethyl- to dioctyltin have been examined; most of them inhibit mitochondrial respiration by preventing the oxidation of keto acids, presumbly via the inhibition of alpha-keto oxidase activity, leading to the accumulation of pyruvate (Aldridge, 1976; Piver, 1973).

Of the trisubstituted compounds, triethyltin compounds have been the most closely examined in relation to the mechanism by which they cause ill effects. Trimethyl- and triethyltin compounds are potent inhibitors of oxidative phosphorylation in the mitochondria for which these compounds have a high binding affinity (Aldridge, 1958; Aldridge & Street, 1964, 1970, 1971). In a later study by Aldridge (1976), triorganotin compounds were stated to derange mitochondrial function in 3 different ways, namely by secondary responses caused by discharge of a hydroxyl-chloride gradient across mitochondrial membranes, by interaction with the basic energy conservation system involved in the synthesis of ATP, and by an interaction with mitochondrial membranes causing swelling and disruption.

Using a ^{113}Sn-labelled triethyltin compound, Rose (1969) identified a pair of histidine residues on guineapig liver mitochondria as the binding site for triethyl compounds. Thus, one molecule of rat haemoglobin binds 2 molecules of triethyltin; the binding sites are located in the globin.

Stockdale et al. (1970) showed that the order of effectiveness for the organotin compounds in inhibiting coupled respiration was tributyl > tripropyl > triphenyl > trimethyl. Two separate effects were suggested: *(a)* an oligomycin-like inhibition of coupled phosphorylation; and *(b)* an alteration of hydroxide exchange across lipid membranes producing uncoupling, swelling, and reduction of intramitochondrial substrate and phosphate concentrations followed by structural damage.

7.2.14 Effective doses and dose rates

7.2.14.1 *Lethal doses*

The median lethal doses for some mono-, di-, tri-, and tetrasubstituted organotin compounds are given in Tables 13, 14, 15, and 16.[a]

Trimethyl and triethyltin compounds, administered orally, are more toxic than the higher homologues of the trialkyltin group. The oral toxicity diminishes progressively from tripropyltin to trioctyltin compounds (Barnes & Stoner, 1958). This is probably because of poorer absorption of higher trialkyltin compounds from the gastrointestinal tract, as judged from the difference between the oral and intraperitoneal toxicity of higher trisubstituted alkyltin compounds (Stoner et al., 1955). The intraperitoneal toxicity of various trialkyltin compounds does not differ to any large extent. In rats, Stoner et al. (1955) found that triethyltin sulfate was equally toxic after intravenous, intraperitoneal, and oral administration. The lethal dose per kg body weight given intravenously or intraperitoneally was 10 mg/kg, causing death within 4—5 days. A dose of 40 mg/kg body weight killed the rat in 2 h. The oral LD_{50} for tricyclohexyltin hydroxide (Table 15) for most species, appears to be between 100 and 1000 mg/kg body weight, while the intraperitoneal and intravenous routes yield LD_{50} values below 20 mg/kg. The difference between oral and parenteral toxicity reflects the poor absorption of tricyclohexyltin hydroxide from the gastrointestinal tract.

In rats and dogs fed with tricyclohexyltin hydroxide, the metabolites, dicyclohexyltin oxide and traces of cyclohexylstannoic acid have been identified. Dicyclohexyltin oxide and cyclohexylstannoic acid also constitute a small portion of the residues on fruit as a result of photodecomposition (FAO/WHO, 1971, pp 522, 534). The median lethal dose of these metabolites may therefore be of interest. The LD_{50} of dicyclohexyltin oxide in the rat appears to be about 350 mg/kg body weight, whereas the LD_{50} value of cyclohexylstannoic acid in this species is probably ten times higher (FAO/WHO, 1971, pp. 524—526).

Only a small amount of data is available concering the acute toxicity of tetraalkyltin compounds. The median lethal dose for tetraethyltin is given in Table 16.

---

[a] Additional data on the toxicity of organotin compounds may be found in a proposal for the United Nations classification and hazard grouping of organotin compounds submitted by the expert from the Netherlands to the Committee of Experts on the Transport of Dangerous Goods of the United Nations Economic and Social Council (EN/CN.2/CONF.5/R.607 of November 1976).

Table 13. Acute oral toxicity of some mono-organotin compounds in various animal species

Trivial name of compound	Medium lethal dose (LD_{50}, mg/kg)	Species	Reference
butylstannoic acid	> 6000	mouse	Pelikán & Černý (1970b)
butyltin trichloride	1400	mouse	
	2140	rat	Marhold (1972)
butyltin-*S*,*S*′*S*″-tris(isooethylmercaptoacetate)	1520	mouse	Pelikán & Černý (1970b)
octyltin trichloride	4600	mouse	Klimmer (1969)
octyltin-*S*,*S*′,*S*″-tris(2-ethylexylmercaptoacetate)	1500	rat (male)	Pelikán & Černý (1970a)

Table 14. Acute toxicity of some diorganotin compounds in various animals

Trivial name of compound	Medium lethal dose (LD_{50}, mg/kg)		Species	Reference
dibutylin di(2-ethylhexoate)	200	(oral)	rat (male)	Calley et al. (1967)
dibutyltin di(butyl maleate)	120	(oral)	rat	Klimmer (1969)
dibutyltin di(nonylmaleate)	170	(oral)		
dibutyltin dichloride	100	(oral)	rat (male)	Klimmer (1969)
	182	(oral)	rat (male)	Mazaev et al. (1971)
	112	(oral)	rat (female)	
	35	(oral)	mouse	
	190	(oral)	guineapig	
	150	(oral)	rat	Marhold (1972)
dibutyltin-*S*,*S*′-bis(2-ethylhexylmercaptoacetate)	150	(oral)	rat (male)	Klimmer (1969)
	175	(oral)		Klimmer (1969)
dibutyltin dilaurate	243	(oral)	rat	Marhold (1972)
	45	(oral)	rat	Marhold (1972)
dibutyltin oxide	520	(oral)	rat (male)	Klimmer (1969)
	39.9	(i.p.)[a]	rat (female)	Robinson (1969)
	24	(oral)	mouse	Mazaev & Šlepnina (1973)
dibutyltin sulfide	145	(oral)	rat (male)	
	150	(oral)	rabbit	
	180	(oral)	rat (female)	
	800	(i.p.)	rat (female)	Robinson (1969)
	> 800	(i.p.)	rat (female)	Robinson (1969)
dioctyltin acetate	2030	(oral)	rat	Klimmer (1969)
diocyltin dibutylmaleate	3750	(oral)	mouse	Pelikan et al. (1970)
dioctyltin-*S*,*S*′-bis(butylmercaptoacetate)	1140	(oral)	white mouse	Pelikán & Černý (1970a)
dioctyltin bis(dodecylmercaptide)	4000	(oral)	white mouse	
dioctyltin bis(2-ethylhexylmaleate)	2760/2750	(oral)	rat (male)	Klimmer (1969)
dioctyltin-*S*,*S*′-bis(2-ethylhexylmercaptoacetate)	2010	(oral)	white mouse	Pelikán & Černý (1970a)
dioctyltin-*S*,*S*′-bis(laurylmercaptoacetate)	3700	(oral)	rat (male)	Klimmer (1969)
dioctyltin-*S*,*S*′-(1,4-butanediol-bis-mercaptoacetate)	2950	(oral)	rat (male)	
dioctyltin dichloride	5500/8500	(oral)	rat (male)	
dioctyltin dilaurate	6450	(oral)	rat (male)	
	800	(i.p.)	rat (female)	Robinson (1969)
dioctyltin di(1,2-propyleneglycolmaleate)	4775	(oral)	rat (male)	Klimmer (1969)
dioctyltin-*S*,*S*′-(ethyleneglycol-bis-dimercaptoacetate)	880	(oral)	rat (male)	
dioctyltin maleate	4500	(oral)	rat (male)	
dioctyltin β-mercaptopropanoate	1850/2050	(oral)	rat (male)	
dioctyltin oxide	2500	(oral)	rat (male)	
dioctyltin mercaptoacetate	945	(oral)	rat (male)	

[a] intraperitoneal

Table 15. Acute toxicity of some triorganotin compounds in various animals

Trivial name of compound	Medium lethal dose (LD_{50}, mg/kg)	Species	Reference
2-trichloro-1-(butine-1'-oxide)-1-triethylstannyloxy)ethane	9.8 (i.p.)	rat	Ignatjeva (1968)
	9.6 (i.p.)	mouse	Ignatjeva (1968)
triethylstannylmethyl-(1-propynyl)formal	10.7 (i.p.)	rat	Ignatjeva (1968)
triethylstannylphenylacetylene	9.9 (i.p.)	mouse	Ignatjeva (1968)
	9.1 (i.p.)	rat	Ignatjeva (1968)
1-triethylstannyl-3-trimethylsiloxi-1-propyne	11.8 (i.p.)	mouse	Ignatjeva (1968)
	11.4 (i.p.)	rat	Ignatjeva (1968)
triethyltin acetate	4 (oral)	rat (female)	Stoner (1966)
triethyltin chloride	5 (i.p.)	rat (female)	Robinson (1969)
triethyltin sulfate	5.7 (i.p.)	rat (male)	Stoner (1966)
	5.3 (i.p.)	guineapig	Stoner (1966)
tributyltin acetate	46 (oral)	white mouse	Pelikán & Černý (1968a)
	99 (oral)		J. Pharm. Sci. (1967)
	133 (oral)	rat (male)	Klimmer (1969)
tributyltin benzoate	108 (oral)	white mouse	Pelikán & Černý (1968a)
	132 (oral)	rat	Arzneimittelforsch. (1969)
tributyltin chloride	117 (oral)	white mouse	Pelikán & Černý (1968a)
	129 (oral)	rat	Marhold (1972)
tributyltin laurate	180 (oral)	white mouse	Pelikán & Černý (1968a)
tributyltin oleate	195 (oral)	rat	Klimmer (1969)
tributyltin salicylate	137 (oral)	rat (male)	Klimmer (1969)
bis(tributyltin oxide)	112—132 (oral)	rat (male)	Klimmer (1969)
	180 (oral)	rat	Truhaut et al. (1976)
	234 (oral)	rat	Sheldon (1975)
	194 (oral, aqueous solution)	rat (male)	Elsea & Paynter (1958)
	148 (oral, oil solution)	rat (male)	Elsea & Paynter (1958)
	11.7 (dermal)	albino rabbit	Elsea & Paynter (1958)
	10.0 (oral)	white mouse	Pelikán & Černý (1969)
trihexyltin acetate	1000 (oral)	rat	Barnes & Stoner (1958)
trioctyltin chloride	10 000 (oral)	rat (male)	Klimmer (1969)
triphenyltin acetate	21 (oral)	guineapig	Klimmer (1963)
	24 (oral)	guineapig	Kimbrough (1976)
	136 (oral)	rat	Klimmer (1963)
	81 (oral)	mouse (male)	FAO/WHO (1971)
	7.9 (i.p.)	mouse (male)	Stoner (1966)
	136 (oral)	rat (male)	Klimmer (1964)
	491 (oral)	rat (female)	Stoner (1966)
	450 (dermal)	rat (male)	Klimmer (1963)
	8.5 (i.p.)	rat (female)	Stoner (1966)
	11.9 (i.p.)	rat (female)	Stoner (1966)
	13.2 (i.p.)	rat (male)	Klimmer (1964)
	21 (oral)	guineapig (male)	Klimmer (1964)
	3.7 (i.p.)	guineapig (male)	Stoner (1966)
	5.3 (i.p.)	guineapig (male)	Klimmer (1964)
	30—50 (oral)	rabbit (male)	Klimmer (1964)
triphenyltin chloride	80 (oral)	rat (male)	FAO/WHO (1971)
	135 (oral)	rat (female)	FAO/WHO (1971)
triphenyltin hydroxide	245 (oral)	mouse (male)	FAO/WHO (1971)
	209 (oral)	mouse (female)	FAO/WHO (1971)
	240 (oral)	rat (male)	Gaines & Kimbrough (1968)
	360 (oral)	rat (female)	Gaines & Kimbrough (1968)

Table 15 (contd)

Trivial name of compound	Medium lethal dose (LD_{50}, mg/kg)	Species	Reference
	27.1 (oral)	guineapig (male)	FAO/WHO (1971)
	31.1 (oral)	guineapig (female)	FAO/WHO (1971)
	171 (oral)	rat (male)	Marks et al. (1969)
	268 (oral)	rat (female)	Marks et al. (1969)
tricyclohexyltin hydroxide	710[a] (oral)	mouse-peromyscus	FAO/WHO (1971)
	1070[a] (oral)	mouse-swiss white	FAO/WHO (1971)
	540 (oral)	rat	FAO/WHO (1971)
	13 (i.p.)	rat	FAO/WHO (1971)
	780 (oral)	guineapig	FAO/WHO (1971)
	9 (i.p.)	guineapig	FAO/WHO (1971)
	500—1000 (oral)	rabbit	FAO/WHO (1971)
	> 126 (i.p.)	rabbit	FAO/WHO (1971)
	150[a] (oral)	sheep	Johnson et al. (1975)
	14 (i.v.)[c]	dog	FAO/WHO (1971)
	6 (i.v.)	cat	FAO/WHO (1971)

[a] approximate lethal dose
[b] intraperitoneal
[c] intravenous

Table 16. Acute oral toxicity of some tetraorganotin compounds in various animal species

Trivial name of compound	Medium lethal dose (LD_{50}, mg/kg)	Species	Reference
tetraethyltin	40.0	mouse	Skačkova (1967)
	15.0	rat	
	40.0	guineapig	
	7.0	rabbit	
	40	mouse	Mazaev et al. (1971)
	9.0	rat	
	40.0	guineapig	
	7.0	rabbit	
tetrabutyltin	6000	rat	Skačkova (1967)

7.2.14.2 *Minimum effective and no-observed-effect doses*

When dibutyltin dichloride was fed to rats for 90 days at dietary levels of 10, 20, 40, and 80 mg/kg, the no-observed effect level was 40 mg/kg, equivalent to 2 mg/kg body weight per day (Gaunt et al., 1968). When dibutyltin was administered to rats for 6 months, the no-observed-effect level was 20 mg/kg diet (Barnes & Stoner, 1958). Dibutyltin sulfide, administered to rats at oral doses of 1.0, 0.1, 0.01, and 0.001 mg/kg body weight, per day for 7 months, was tolerated without any detected adverse effects up to a dose of 0.01 mg/kg per day but the higher levels of 0.1 and 1.0 mg/kg per day caused intoxication (Mazaev & Korolev, 1969). By comparison, the toxicity of dioctyltin compounds was relatively low (Klimmer, 1969). Thus, dioctyltin did not produce any injury in rats, mice, or guineapigs at oral doses up to 400 mg/kg body weight per day when given for 3—4 successive days, nor were there any ill effects when it was added to the daily diet of rats for 4 months at a rate of 200 mg/kg (Barnes & Stoner, 1958).

Data on effective doses of trisubstituted organotin compounds are available only for triphenyltin derivatives and tricyclohexyltin hydroxide. Triphenyltin acetate was administered, by gavage, to groups of rats at doses equivalent to dietary levels of 5, 10, 25, and 50 mg/kg for up to 17 days. While no adverse effects were found at the 25 mg/kg level, the 50 mg/kg level resulted in the death of animals, mainly as a result of infection (Klimmer, 1964). In a 2-year study on rats fed diets containing concentrations of triphenyltin hydroxide of up to 10 mg/kg, the no-observed-effect level was about 2 mg/kg (equivalent to about 0.1 mg/kg body weight per day) (FAO/WHO, 1971, pp. 341—342). In another 2-year study in which groups of guineapigs were fed diets containing concentrations of triphenyltin acetate, up to 200 mg/kg, the no-observed-effect level was 5 mg/kg. At higher levels there was a dose-related increase in mortality and the occurrence of other effects such as inhibition of growth rate and fatty degeneration of the liver and heart (FAO/WHO, 1971, p. 341). With respect to tricyclohexyltin hydroxide, no toxic effects were noted in rats receiving 12.5 mg/kg body weight daily for 19 days, whereas a rate of 25 mg/kg per day caused toxic effects (FAO/WHO, 1971, pp. 527—528). Rats were fed on diets providing up to 12 mg/kg body weight per day of tricyclohexyltin hydroxide in a 2-year study; the no-observed-effect level was 3 mg/kg body weight per day (FAO/WHO, 1971, p. 528). When dogs were fed tricyclohexyltin hydroxide in the diet at concentrations of up to 12 mg/kg body weight per day for 6—24 months, many dogs refused the 12 mg/kg per day diet, causing weight loss and death from starvation. The no-observed-effect level was found to be about 0.75 mg/kg per day (FAO/WHO, 1971, pp. 526—527). Johnson et al. (1975) also reported the effects of thoroughly wetting the skin of yearling cattle, goats, and sheep with suspensions of tricyclohexyltin hydroxide. They found that cattle tolerated suspensions up to a concentration of 5 g/litre while anorexia was registered in some animals after spraying with a suspension of 10 g/litre. Yearling goats and sheep tolerated suspensions of up to 1 g/litre. Transitory anorexia and eye irritation developed after application of a suspension of 20 g/litre to the skin of goats.

Few data are available on the effective dose levels of tetra-substituted organotin compounds. Intravenous administration of tetraethyltin at 25 mg/kg body weight produced a slight increase in respiratory rate and vasodilatation as immediate effects, but, 1.5—2 h later, prostration with muscular weakness occurred. (Stoner et al., 1955).

8. EFFECTS ON MAN

There are comparatively few clinical observations and epidemiological data concerning the effects of inorganic tin compounds on man and even fewer on the effects of organotin compounds.

8.1 Inorganic Tin Compounds

8.1.1 Acute poisoning

Some episodes of acute poisoning have been reported, mainly in association with the ingestion of fruit juices containing high concentrations of tin. The major symptoms and signs noted were nausea, vomiting, diarrhoea, fatigue, and headache. The concentrations of tin in the products thought to have been associated with the incidents were uncertain in many cases, but were probably in the range of 300—500 mg/kg (Horio et al., 1967). Orange and apple juice, containing tin concentrations of 250—385 mg/kg were suspected of causing one incident (Benoy et al., 1971). Nausea, vomiting, and diarrhoea were recorded in individuals consuming peach preserves that contained a tin concentration of 563 mg/kg, zinc at 1.5 mg/kg, cadmium at 0.1 mg/kg, lead at 0.16 mg/kg, copper at 1 mg/kg, nitrate at 93 mg/kg, nitrite at 1.7 mg/kg, and chloride at 115 mg/kg (Nehring, 1972). Acute gastroenteritis followed ingestion of a fruit punch stored in a tin can and containing a tin concentration of 2000 mg/litre. First symptoms occurred after 1—2 h, the earliest and commonest being bloatedness, followed by severe nausea, stomach cramps, vomiting and, in one-third of the patients, mild diarrhoea (Warburton et al., 1962). Other similar outbreaks have involved canned cherries (Luff & Metcalf, 1890), asparagus, herring (Schryver, 1909), and apricots (Savage, 1939) with tin concentrations ranging from 300 to about 1000 mg/kg.

An incident of acute intoxication after the ingestion of canned peaches was reported by Svensson (1975). It was unique in that about 110 participants in a meeting received nothing else to eat or drink, except for the peaches. The report was based on questionnaires received from 85 persons, 76 (89%) of whom had fallen ill. About half of these had developed symptoms such as nausea, vomiting, and diarrhoea within 1 h. The majority became symptom-free after 24 h. The fruit contained a mean tin concentration of 533 mg/kg (range 413—597 mg/kg) and the juice 369 mg/kg (range 298—405 mg/kg). Two out of 7 persons who had consumed only one quarter of the contents fell ill. The calculated intake of tin for these persons was 50 mg.

In another report, canned tomato juice was associated with 113 cases of acute gastroenteritis (Barker & Runte, 1972). In a small number of these cases, superficial erosion of the mucosa of the mouth was observed, while abdominal distension and cramps, and diarrhoea were quite common. The cans containing the juice showed complete corrosion of tin linings and this yielded a product containing a tin concentration of approximately 400 mg/kg. The crops of tomatoes, used to make this particular juice, had been grown in soil excessively treated with nitrate fertilizer and contained high levels of nitrate; complete corrosion of the lining of the cans occurred within approximately 6 months.

Five human volunteers did not show any toxic signs after drinking fruit juices containing about 500 or 730 mg/kg of tin, but all had some gastrointestinal disturbance after drinking 5—7 ml/kg body weight of fruit juice containing a tin concentration of about 1400 mg/litre (Benoy et al., 1971). There was no evidence to indicate that the effects were due to the absorption of tin, the likeliest cause being local irritation of the mucous membranes of the alimentary tract.

8.1.2 Prolonged exposure

8.1.2.1 *Effects of inhalation*

The only available information on exposure by inhalation pertains to pneumoconiosis caused by the inhalation of tin(IV) oxide. This benign condition is termed stannosis.

More than 200 cases of stannosis have been described (Bartak et al., 1948; Cutter et al., 1949; Dundon & Hughes, 1950; Oyanguren et al., 1958; Pendergrass & Pryde, 1948; Robertson & Whitaker, 1957; Schuler et al., 1958). The relative importance of exposure to tin fumes in the etiology of this disorder was emphasized by Dundon & Hughes (1950). The significance of the quantity of dust and the duration of exposure were stressed by Robertson & Whitaker (1957), who reviewed 121 cases.

Pendergrass & Pryde (1948) noted stannosis in a man who, for 15 years, had been bagging tin oxide material containing 96.5% tin(IV) oxide and small amounts of aluminium, iron, and sodium but not silica. Radiography revealed small dense shadows (denser than those of silicosis) resembling those of barytosis in both lungs. Bartak et al., (1948) reported a similar case in a workman who had suffered from asthma for many years and who attended a furnace in which metallic tin was burnt to produce tin(IV) oxide. Necropsy of this man, who died from gastric carcinoma, revealed deposits of tin(IV) oxide in the lungs, lymph glands, liver, and spleen. Six other workmen with a similar radiographic appearance of the lungs, did not have any symptoms of asthma or signs of pulmonary dysfunction. Similar radiological findings were reported by Cutter et al. (1949) in 2 cases with nodules 1—2 mm in diameter unaccompanied by evidence of pulmonary dysfunction. Both had been working in a tin recovery department for 20 years.

Oyanguren et al., (1958) reported 10 cases of stannosis in workers involved in a process where tin ore concentrates were reduced to metal. The workers were exposed to dust with a high tin and a low silica content and to fumes rich in tin. None of the exposed workers had any disability and the vital capacity, maximal breathing and resting minute volume, and respiratory reserve were normal, as were the findings on the blood and urine. Radiological examination of these 10 workers showed that the first alteration appeared to be increase of bronchovascular markings, with hilar thickening in the early stages; later, well-defined nodular elements appeared first in the middle third of the right and then of the left lung. These appeared to progress to the rest of the lung fields, and with continuous exposure, accumulation of tin(IV) oxide gave a metallic density to a part of the nodules, which subsequently acquired an appearance of lipoid droplets. In the later stages, the bronchovascular markings disappeared as the density of the shadows increased. The radiographic changes appeared after some 3—5 years of exposure (Schuler et al., 1958). Hlebnikova (1957) made a survey over a number of years of workers, who were exposed to condensation aerosols, formed during the smelting of tin and consisting mainly of tin(IV) oxide. The free silica concentration in the aerosols was not more than 1% and the total silica did not exceed 3%. The total dust concentration in air varied between 3 and 70 mg/m^3. Workers developed pneumoconiosis after 6 to 8 years of working at the smelter. The authors described 45 cases of pneumoconiosis in workers, who were employed at the smelter from 6 to 20 years, Six of them were reported to have second degree pneumoconiosis. X-ray examination showed small spotty shadows, but there were no other symptoms or signs. The opacities seen on the radiographs were thought to be due to the accumulation of tin-containing dust and the development of connective tissue, this was confirmed by experiments on rats exposed to tin(II) oxide. No cases of pneumoconiosis were observed in 10 years, after the dust concentration had been reduced to 10 mg/m^3. An important feature of stannosis is that fibrosis of the lung does not develop, providing that other agents such as silica are not present.

8.1.2.2 *Effects of ingestion*

Packaged military rations were fed to 9 young male adult volunteers for successive 24-day periods. The average tin content of a control fresh diet was 13 mg/kg (in dry solids), while C-rations stored at 1°C contained a tin concentration of 33 mg/kg, and rations stored at 37 °C contained 204 mg/kg. All the tin ingested was accounted for by faecal elimination. No toxic effects were noted (Calloway & McMullen, 1966). Dack (1955) reported a study in which 4 subjects ate canned pumpkin containing a tin concentration of about 380—480 mg/kg and canned asparagus containing a concentration of about 360 mg/kg, for 6 days, with no apparent illness. In an earlier study, Mamontova (1940) did not observe any toxic manifestations in 4 volunteers who, for a period of 30 days, consumed 250 g per day of canned fruit containing 212—250 mg of tin.

8.2 Organotin Compounds

8.2.1 Local effects

Toxic lesions among laboratory and process workers handling di- and tributyltin compounds were reported by Lyle (1958). Most of the lesions were typical acute skinburns, caused by the colourless di- or tributyltin chlorides, which can come into contact with the skin without exposed workers being aware of it. This type of lesion developed 1 to 8 h following exposure. When the substance was washed off immediately, no lesion developed. A more diffuse, but less rapidly healing lesion, was caused by contact with clothes that had been moistened by vapour or liquid compounds. This subacute irritation was characterized by itching, affecting mostly the skin of the lower abdomen, thighs, and groin. On examination, an easily distinguishable erythematous eruption, was noted. Cessation of contact with the compound was followed by rapid healing. Following contact with the eye, lachrymation and severe suffusion of the conjunctivae appeared within a few minutes and persisted for 4 days. Immediate lavage of the eye did not prevent the development of the signs.

Lyle (1958) also studied the skin lesions induced by butyltin compounds by applying various compounds on the skin of the back of the hands of 5 volunteers. Skin lesions could be produced by a single application of dibutyltin dichloride, and of the chloride, acetate, and oxide of tributyltin, while the diacetate, dilaurate, oxide, and maleate of dibutyltin and also tetrabutyltin failed to produce any lesions. No visible changes occurred for 2 to 3 h after application, although a swelling of the mouth of the hair follicles was noticeable. The follicular inflammation progressed during the following 8 h with only slight visible irritation of the skin between the openings of the follicles. On the second day, sterile pustules developed over the follicular openings, but remained small during the next 3 to 4 days. After one week, the lesions had practically disappeared.

Symptoms experienced by female spray-painters working with a latex paint, to which was added a fungicidal solution containing 20% bis(tributyltin) oxide, ethylene oxide, ethanol, and water, were described by Landa et al. (1973). The symptoms were reported to appear "immediately" after the start of spraying and were experienced by all the women engaged in the work. The first sensations were irritation of the nasal mucosa and the conjunctivae. The exposure continued for another fortnight during which the symptoms and signs became more severe, and included bleeding from the nose and mucous discharges. An otorhinolaryngological examination revealed rhinitis with distinct hyperaemia and haemorrhages of the nasal septum. The workers reported that the symptoms were less severe during weekends. When the addition of the fungicidal solution was abandoned, the symptoms disappeared. One year later, identical symptoms were reported by 4 spray-painters using latex paint. An inquiry disclosed that the manufacturer had begun to use bis(tributyltin) oxide as a fungicidal additive since the banning of the use of mercury. The authors considered that the trialkyltin compound was responsible for the symptoms. Measurements performed later indicated that the concentrations of tin in the breathing were below 0. 05 mg/m^3 air.

Another description of local effects seen in workers handling triphenyltin acetate was given by Markićević & Turko (1967). These workers were engaged in the formulation of a 20% solution of a fungicide. During the hottest summer days, subjects working in the dustiest places developed conjunctival irritation and irritation of the mucous membranes of the upper respiratory tract as well as of the skin, the hands and scrotum being particularly affected. Effects on the central nervous system were not registered. The signs disappeared rapidly after termination of exposure.

A wettable powder formulation containing 50% tricyclohexyltin hydroxide was examined for its irritation and sensitization potential. No adverse reactions were observed in 53 females after sensitization applications and a challenge application 20 days later of 0.5 ml of an emulsion (10 g/litre). Tricyclohexyltin hydroxide was reported not to be dermally irritating at a concentration of about 0.01 mg/kg body weight (FAO/WHO, 1971)

8.2.2 Systemic effects

8.2.2.1 *Effects of Dermal exposure*

In a recent review (NIOSH, 1976), a fatal case was described in which a 29-year-old woman was accidentally drenched in a slurry containing triphenyltin chloride, diphenyltin dichloride, hexane, and other unidentified compounds at a temperature of 79.4°C. On arrival at hospital, first-degree thermal burns covered 10% of her body. Second- and third-degree burns with 80—85% desquamated skin developed 12 h later. Death from renal failure occurred 12 days after the accident. However, the agent responsible for her symptoms and signs and for the death could not be identified from the data available.

Mijatović (1972) reported a case of systemic intoxication resulting from dermal contact with a compound containing triphenyltin acetate (60%) and manganese dithiocarbamate (15%). The subject, engaged in agricultural aviation, spilt the compound on his hands and chest while filling the aeroplane. The skin on his chest and abdomen was indured after 3 h and vesicles developed the following day. He experienced headache, nausea, epigastric pain, and general weakness. Clinically the liver transaminases (SGPT) were elevated, reaching a maximum one month later. Two months later the transaminases were returning to normal, but the patient had pains over the liver, which was tender and enlarged. The liver damage persisted for 2 years and the case was labelled chronic hepatitis. No liver biopsy was performed.

8.2.2.2 *Effects of inhalation*

Acute intoxication caused by the inhalation of triphenyltin compounds has been reported in some instances. Three cases occurred in farmers treating beetroot plants with triphenyltin acetate which was inhaled (Guardascione & di Bosco, 1967). The symptoms started from a few minutes to about 2 h after the first exposure. After 2 hours of spraying, one subject felt a general malaise and severe headache and eventually lost consciousness. On admission to the hospital he had a diffuse tremor and a slightly depressed sensorium. All laboratory investigations during the 2-week hospitalization were normal. When mixing triphenyl tin acetate powder into a solution, a second subject experienced repeated flushes, nausea, and shortness of breath, which started only a few minutes after inhalation. During the 6-day hospital surveillance, the only pathological finding was glucosuria. A third subject suffered from severe headache, strong nausea, and epigastric pains. However, laboratory investigations were normal. The headache persisted for 2 days while the epigastric pains and the nausea disappeared after one day. All 3 subjects recovered completely.

Horáček & Demčik (1970) reported a group poisoning involving 2 pilots and 3 mechanics following the spraying of a formulation comprising 60% of triphenyltin acetate and 15% of manganese dithiocarbamate. Protective measures were found to be deficient and food was consumed with unwashed hands; thus, exposure by both inhalation and ingestion could have occurred. Furthermore, other pesticides containing copper, zinc, DDT, and organophosphates were handled during exposure to the formulation, but the authors considered these substances unlikely to be responsible for the symptoms experienced. Exposure to the formulation continued for 2 weeks before symptoms were recognized and continued for 2 further weeks. One pilot experienced gastric pain, diarrhoea, and dryness of the mouth with severe thirst that was not realieved by drinking. He also felt pressure over the chest and slight shortness of breath. Blurred vision was experienced after one week of exposure. Clinically hepatomegaly with a painful liver was noted. Liver transaminases (SGPT) were elevated and the highest recorded value was obtained 6 weeks later. Hyperglycaemia and glycosuria were also present. Eight weeks afterwards, liver biopsy revealed marked diffuse steathosis without any detected necrosis. Monthly follow-ups showed that hepatomegaly and steathosis persisted for one year. The second pilot suffered from heartburn, diarrhoea, and vision disturbances. Clinically a hepatomegaly and a slight hyperglycaemia were registered. The symptoms persisted for 4 weeks and the glycosuria and hepatomegaly for 6 weeks. Recovery was complete. The mechanics had symptoms that were less severe, comprising diarrhoea, headache, eye pains, blurred vision, epigastric pain, and thirst.

A worker engaged in the manufacture of butyltin compounds was reported to suffer from a reduced sense of smell (Akatsuka et al., 1959). It was first observed after an exposure period of 16 months, and a further deterioration of the olfactory sense was established during the following 8 months. The state persisted without any noted improvement for 2 years. Other reported symptoms were headaches in the occipital region, nasal haemorrhages, lassitude, and a feeling of stiffness in the shoulders.

In four cases of acute poisoning due to exposure to organotin vapours, patients were reported to have suffered from such symptoms as vertigo, headaches, nausea and vomiting, and visual disturbances. Clinically, stasis of the papilla was found and all patients displayed pathological findings on the electroencephalograms. These were reversible in 7—25 days, and all cases recovered clinically. The organotin compound or compounds responsible for the intoxications were not identified (Prüll & Rompel, 1976).

8.2.2.3 *Effects of ingestion*

A most serious episode involving organotin poisoning occurred in 1954 due to oral administration of a proprietary preparation used for the treatment of furonculosis, osteomyelitis, anthrax, and acne. The drug was responsible for about 100 deaths and a total of about 210 intoxications. These numbers vary to some extent depending on the source of information: Alajouanine et al., (1958) reported a total of 210 deaths whereas another source quotes 102 deaths and at least 100 persons permanently affected (*Br. med. J.*, 1958). A total of about 400 000 capsules was sold (*Br. med. J.*, 1958), and about 1000 persons were believed to have taken the drug (Barnes & Stoner, 1959). However, the capsules consumed by the intoxicated subjects amounted only to about 7% of the lot distributed (*Br. med. J.*, 1958), implying that the majority of the capsules were consumed without any known adverse effects. The main ingredients of the preparation were diethyltin diiodide (15 mg/capsule) and linoleic acid (100 mg/capsule). It was suggested that ethyltin triiodide, triethyltin iodide, or tetraethyltin could have been present as impurities because of deficiencies in the manufacturing process or as metabolites formed under the influence of various physical factors (Rondepierre et al., 1958). A theory that diethyltin diiodide would have reacted with the isolinoleic component producing tetraethyltin was also presented (Lecoq, 1954). The symptoms and clinical findings described in victims by Alajouanine et al. (1958) seem to favour the hypothesis of a trialkyltin compound being the causal agent, probably acting synergistically with other constituents. The clinical data of 201 cases including 98 deaths were reviewed by Alajouanine et al. (1958). The dominating symptom, reported in about 98% of the cases, was a diffuse headache, sometimes intolerably severe, and appearing a few days after medication was started. Nausea and vomiting occurred in 73%, visual disturbances, mainly photophobia, but also double vision, colour-vision disturbances and, in a few cases, blindness were recorded in 33% of cases. Ophthalmoscopic findings included congestion, papilloedema (Druault-Toufesco, 1955), and in some cases, papillary stasis (Gayral et al., 1958; Pesme, 1955). Frequent symptoms and signs were urinary incontinence, vertigo, loss of weight, and abdominal pains. Absence of fever and a tendency towards hypothermia were also noted. Psychological disturbances or stupor were reported in 70% of the cases. Other findings were meningeal irritation, somnolence, insomnia, convulsions, constipation, and bradycardia. Sometimes electroencephalograms were altered but did not suggest any localized lesion. Death occurred during coma or from respiratory or cardiac failure and in some cases during convulsions. It is probable that most of the symptoms and signs could be attributed to a cerebral oedema, the occurrence of which was established at autopsies and decompressive surgery (Cossa et al., 1958, Fontan et al., 1958). This cerebral oedema of the white matter appeared to be very similar to that produced experimentally by administration of a proprietary preparation containing an organotin compound, to mice and monkeys (Gruner, 1958). Macroscopically, the brain was oedematic but the microscopic findings were minor.

It has been reported that only 10 out of 103 subjects, who survived, recovered completely: in the remainder, symptoms such as headaches and asthenia persisted for at least 4 years (Barnes & Stoner, 1959). Information on later, follow-up studies is not available. The lethal dose of the preparation was in some instances only about 25 capsules taken during one week. Ingestion of 3 capsules was enough to cause intoxication in a 9-year-old child (Fontan et al., 1955).

8.3 Treatment of Poisoning

Dimercaprol has been suggested by Stoner et al. (1945) to be an effective antidote for dialkyltin poisoning. It has also been reported to completely prevent the accumulation of alpha-keto acids produced by dialkyltin compounds (Barnes & Stoner, 1959). Although dimercaprol protected rats against the general toxic effects of these compounds, it did not have any effect on the response to triethyltin compounds. This seems to be due to the fact that dialkyltin compounds, at least up to the dihexyl derivatives, react readily with sulfhydryl groups and the trialkyltin compounds do not (Barnes & Magos, 1968),

Studer et al. (1973) reported that steroid therapy (dexamethasone) appeared to diminish mortality and the severity of brain oedema in rats; there also appeared to be significant decreases in brain, liver, and blood levels of triethyltin bromide. The authors suggested that the beneficial effects of this steroid therapy might be partly due to enhanced excretion or catabolism of triethyltin bromide.

Surgical decompression was considered to be the only treatment that offered any benefit in human cases of cerebral oedema caused by trialkyltin compounds (Alajouanine et al., 1958).

REFERENCES

ADAM, W. B. & HORNER, G. (1937) The tin content of English canned fruits and vegetables. *J. Soc. Chem. Ind.,* **56:** 329—334.

ADAMSON, J. H. (1962) The colorimetric determination of dialkyltin compounds dissolved in fats and olive oil. *Analyst,* **87:** 597—598.

ADCOCK, L. H. & HOPE, W. G. (1970) A method for the determination of tin in the range 0.2 to 1.6 μg, and its application to the determination of organotin stabilizer in certain foodstuffs. *Analyst,* **95:** 868—874.

AKAGI, H. & SAKAGAMI, Y. (1971) [On degradation of organotin compounds by ultraviolet ray.] *Rep. Public Health Inst.,* **20:** 1—4 (in Japanese).

AKAGI, H., FUJITA, M., & SAKAGAMI, Y. (1972) [Thin layer chromatography of inorganotin and organotin compounds in foods.] *J. food Hyg.,* **13:** 85—88 (in Japanese).

AKATSUKA, K., MIYAZAWA, J., IGARASHI, I., MORISHITA, M., HANDA, A., KAWAME, N., IWAMOTO, I., MORITO, F., MURAYAMA, K., NAKANO, S., YANAGIBASHI, H., NAGASAKI, T., KOTANI, Y., MATSUTANI, W., FUKUDA, I., & IYO, T. (1959) [Experimental studies on disturbance of sense of smell due to butyltin compounds.] *J. Tokyo Med. Coll,* **17:** 1393—1402 (in Japanese).

ALAJOUANINE, T., DÉROBERT, L., & THIÉFRY, S. (1958) Etude clinique d'ensemble de 210 cas d'intoxication par les sels organiques d'étain. *Rev. Neurol.,* **98:** 85—96.

ALDRIDGE, W. N. (1958) The biochemistry of organotin compounds; trialkyltins and oxidative phosphorylation. *Biochem. J.,* **69:** 367—376.

ALDRIDGE, W. N. (1976) The influence of organotin compounds on mitochondrial functions. In: Zuckerman, J. J., ed. *Organotin compounds: New chemistry and applications,* Washington, DC, American Chemical Society, pp. 186—196.

ALDRIDGE, W. N. & CREMER, J. E. (1957) Organotin dithizone complexes: The colorimetric determination of diethyltin and triethyltin compounds. *Analyst,* **82:** 37—43.

ALDRIDGE, W. N. & STREET, B. W. (1964) Oxidative phosphorylation: Biochemical effects and properties of trialkyltins. *Biochem. J.,* **91:** 287—297.

ALDRIDGE, W. N. & STREET, B. W. (1970) Oxidative phosphorylation: The specific binding of trimethyltin and triethyltin to rat liver mitochondria. *Biochem. J.,* **118:** 171—179.

ALDRIDGE, W. N. & STREET, B. W. (1971) Oxidative phosphorylation: The relation between the specific binding of trimethyltin and triethyltin to mitochondria and their effects on various mitochondrial functions. *Biochem. J.,* **124:** 221—234.

ALEU, F. P., KATZMAN, R., & TERRY, R. D. (1963) Fine structure and electrolyte analysis of cerebral edema induced by alkyltin intoxication. *J. Neuropathol exp. Neurol.,* **22:** 403—413.

ALLAN, J. E. (1962) Atomic absorption spectrophotometry absorption lines and detection limits in the air-acetylene flame. *Spectrochim. Acta,* **18:** 259—263.

AMOS, M. D. & WILLIS, J. B. (1966) Use of high-temperature pre-mixed flames in atomic absorption spectroscopy. *Spectrochim. Acta,* **22:** 1325—1343.

ANALYTICAL METHODS COMMITTEE (1967) The determination of small amounts of tin in organic matter. *Analyst,* **92:** 320—323.

ANDERSON, J. R. A. & LOWY, S. L. (1956) A fluorimetric determination of tin. *Anal. Chim. Acta,* **15:** 246—253.

ASCHER, K. R. S. & NISSIM, S. (1964) Organotin compounds and their potential use in insect control. *World Rev. Pest Control,* **3:** 188—211.

AVTANDILOV, G. G. (1967) [Trace element content in normal and atherosclerotic portions of the human aorta in connection with age.] *Arh. Patol.,* **29:** 40—42 (in Russian).

BARKER, W. H. & RUNTE, V. (1972) Tomato-juice-associated gastroenteritis, Washington and Oregon, 1969. *Am. J. Epidemiol.* **96:** 219—226.

BARNES, J. M. & MAGEE, P. N. (1958) The biliary and hepatic lesion produced experimentally by dibutyltin salts. *J. Pathol. Bacteriol.* **75:** 267—279.

BARNES, J. M. & MAGOS, L. (1968) The toxicology of organometallic compounds. *Organometal. Chem. Rev.,* **3:** 137—150.

BARNES, J. M. & STONER, H. B. (1958) Toxic properties of some dialkyl and trialkyl tin salts. *Br. J. ind. Med.,* **15:** 15—22.

BARNES, J. M. & STONER, H. B. (1959) The toxicology of tin compounds. *Pharmacol. Rev.,* **11:** 211—231.

BARNES, R. D., BULL, A. T., & POLLER, R. C. (1973) Studies on the persistence of the organotin fungicide, fentin acetate (triphenyltin acetate) in the soil and on surfaces exposed to light. *Pestic. Sci.,* **4:** 305—317.

BARTÁK, F., TOMEČKA, M., & TOMIČEK, O. (1948) Stannosis. *Čas. lék. česk.,* **87:** 915—920 (Abstract in: *J. ind. Hyg.,* (1949) **31** (5): 98—99)

BASSINDALE, A. R., EABORN, C., TAYLOR, R., THOMPSON, A. R., WALTON, D. R. M., CRETNEY, J., & WRIGHT, G. J. (1971) Aromatic reactivity 46. Interpretation of substitutent effects in base-catalyzed cleavage of aryltrimethylsilanes and aryltrimethyl stannanes. *J. Chem. Soc. B.,* 1155—1161.

BENNETT, R. L. & SMITH, H. A. (1959) Spectrophotometric determination of tin with phenylfluorone. *Anal. Chem.,* **31:** 1441—1442.

BENOY, C. J., HOOPER, P. A., & SCHNEIDER, R. (1971) The toxicity of tin in canned fruit juices or solid foods. *Food Cosmet. Toxicol.,* **9:** 645—656.

BENS, A. A., GRABOVSKAJA, L. I., & TIHONOVA, I. V. (1976) [*Geochemistry of the environment.*] Moscow, Nedra (in Russian).

BERGNER, K. G. & RÜDT, U. (1958) [On the absorption of lead and tin by fat and food rich in fat in tin plate cans.] *Z. Lebensm. Unters. Forsch.,* **137:** 337—351 (in German).

BERTINE, K. K. & GOLDBERG, E. D. (1971) Fossil fuel combustion and the major sedimentary cycle. *Science,* **173:** 233—235.

BISTON, R., VERSTRAETEN, E., & NANGNIOT, P. (1972) Dosage de l'étain dans les conserves par polarographie oscillographique à intensité imposée: Comparaison avec les methodes colorimetriques au dithiol et à la phenylfluorone. *Anal. Chim. Acta,* **59:** 453—460.

BLAIR, E. H. (1975) Biodegradation of tricyclohexyltin hydroxide. *Environ. Qual. Saf.,* **3** (Suppl.): 406—409.

BLASIUS, M. B., KERKHOFF, S. J., WRIGHT, R. S., & COTHERN, C. R. (1972) Use of X-ray fluorescence to determine trace metals in water resources. *Water Resour. Bull.,* **8:** 704—714.

BOGEN, J. (1973) Trace elements in atmospheric aerosol in the Heidelberg area, measured by instrumental neutron activation analysis. *Atmos. Environ.,* **7:** 1117—1125.

BÖNIG, G. & HEIGENER, H. (1972) [Paper chromatography: Quantitative determination of micro amounts of tin.] *Landwirtsch. Forsch.,* **25:** 378—382 (in German).

BOWEN, H. J. M. (1966) *Trace elements in biochemistry,* New York, Academic Press, pp. 203—204.

BOWEN, V. T. & SUTTON, D. (1951) Comparative studies of mineral constituents of marine sponges. I. The Genera *Dysidea, Chondrilla, Terpios. J. mar. Res.,* **10:** 153—167.

BRAMAN, R. S. & TOMPKINS, M. A. (1979) Separation and determination of nanogram amounts of inorganic tin and methyltin compounds in the environment. *Anal. Chem.* **51:** 12—19.

BRIDGES, J. W., DAVIES, D. S., & WILLIAMS, R. T. (1967) The fate of ethyltin and diethyltin derivates in the rat. *Biochem. J.,* **105:** 1261—1266.

BRINCKMAN, F. E. & IVERSON, W. P. (1975) Marine chemistry in the coastal environment. Paper presented at *169th American Chemical Society Meeting. Philadelphia, PA., April 8—10.*

BRINCKMAN, F. E., BLAIR, W. R., JEWETT, K. L., & IVERSON, W. P. (1977) Application of a liquid chromatograph coupled with a flameless atomic absorption detector for speciation of trace organometallic compounds. *J. Chromatogr. Sci.,* **15:** 493—503.

BRITISH MEDICAL JOURNAL (1958) "Stalinon": A therapeutic disaster. *Br. med. J.,* March 1st, p. 515.

BROWNING, E. (1969) Tin. In: *Toxicity of industrial metals,* 2nd ed., London Butterworths, pp. 323—330.

BRÜGGEMANN, J. & KLIMMER, O. R. (1964) [Nutritional-physiological, analytical and toxicological studies on fungicidal triphenyltin acetate.] *Zentralbl. Vet.-Med. A,* **11:** 1—3 (In German).

BRÜGGEMANN, J., BARTH, K., & NIESAR, K.-H. (1964a) [Experimental studies on the content of triphenyltin acetate residues in turnip leaves, ensilage of turnip leaves, and in excrements of animals fed with these products.] *Zentralbl. Vet.-Med. A,* **11:** 4—19 (in German).

BRÜGGEMANN, J., KLIMMER, O. R., & NIESAR, K.-H. (1964b) [The importance of triphenylin acetate residues in plants and animals from a hygienic-toxicologic standpoint.] *Zentralbl. Vet.-Med. A,* **11:** 40—48 (in German).

BÜRGER, K. (1963) [Thin layer chromatographic method for the detection, identification and separation of traces of mono-, di-, tri- and zerovalent organotin compounds.] *Z. anal. Chem.,* **192:** 280—286 (in German).

BYINGTON, K. H., YEH, R. Y., & FORTE, L. R. (1974) The hemolytic activity of some trialkyltin and triphenyltin compounds. *Toxicol. appl. Pharmacol.,* **27:** 230—240.

CALLEY, D. J., GUESS, W. L., & AUTIAN, J. (1967) Hepatotoxicity of a series of organotin esters. *J. Pharm. Sci.,* **56** (2): 240—243.

CALLOWAY, D. H. & MCMULLEN, J. J. (1966) Fecal excretion of iron and tin by men fed stored canned foods. *Amer. J. clin. Nutr.,* **18:** 1—6.

CALVERY, H. O. (1942) Trace elements in foods. *Food Res.,* **7:** 313—331.

CAPACHO-DELGADO, L. & MANNING, D. C. (1966) Determination of tin by atomic absorption spectroscopy. *Spectrochim. Acta,* **22:** 1505—1513.

CARR, H. G. (1969) Interaction between PVC bottles and liquid foods. *Soc. Plast. Eng.* **25:** 72—74.

CASIDA, J. E., KIMMEL, E. C., HOLM, B., & WIDMARK, G. (1971) Oxidative dealkylation of tetra-, tri-, and dialkyltin and tetra- and trialkyleads by liver microsomes. *Acta chem. Scand.,* **25:** 1497—1499.

CATALA, R., ROYO, I. J., & PRIOMO, E. (1971) Tin content of canned orange juice. *Rev. Agroquim. Tecnol. Aliment.,* **11:** 409—417 *(Chem. Abstr.,* (1972) **77:** 370, Abstract no. 3974 w).

CAUJOLLE, E., LESBRE, M., & MEYNIER, D. (1954) Sur la toxicité du tétraméthylstannane et du tétraéthylstannane. *C. R. Acad. Sci (Paris),* **239:** 556—558.

CENCI, P. & CREMONINI, B. (1969) [Thin-layer chromatography of two organotin pesticides and their behavior in various types of soil.] *Ind. sac. Ital.,* **69:** 313—316 (in Italian).

CHAPMAN, A. H., & PRICE, J. W. (1972) Degradation of triphenyltin acetate by ultraviolet light. *Int. Pest Control,* **14**: 11—12.

CHAPMAN, A. H., DUCKWORTH, M. W., & PRICE, J. W. (1959) The determination of dialkyltin compounds in polyvinylchloride. *Br. Plast.,* February, pp. 78, 87.

CHEFTEL, H. (1967) *L'étain dans les aliments.* Presented at the Fourth Meeting of the FAO/WHO Codex Committee on Food Additives, The Hague, 11—15 September 1967, Rome, Food and Agriculture Organization of the United Nations, 10 pp.

CHISAKA, H., TANIZAKI, Y., & NAGATSUKA, S. (1973) [Determination of tin and poisonous trace elements in canned juices by neutron activation.] *Radioisotopes,* **22**: 247—248 (in Japanese).

CHRISTIAN, G. D. & FELDMAN, F. T. (1970) *Atomic absorption spectrography.* New York, Wiley, pp. 404—406.

CONINE, D. L., MUHLER, J. C., & FORNEY, R. B. (1972) Acute toxicity of sodium pentafluorostannite in rats and mice. *Toxicol. appl. Pharmacol.,* **22**: 303—304.

CONINE, D. L., YUM, M. N., MARTZ, R. C., STOOKEY, G. K., & FORNEY, R. B. (1973) Comparison of the metabolic and renal effects of sodium pentafluorostannite, sodium fluoride, stannous chloride, hydrochloric acid, and starvation in the rat. *Pharmacologist,* **15**: 189.

CONINE, D. L., YUM, M. N., MARTZ, R. C., STOOKEY, G. K., MUHLER, J. C., & FORNEY, R. B. (1975) Toxicity of sodium pentafluorostannite, a new anticariogenic agent. I. Comparison of the acute toxicity of sodium pentafluorostannite, sodium fluoride and stannous chloride in mice and/or rats. *Toxicol. appl. Pharmacol.,* **33**: 21—26.

CONINE, D. L., YUM, M., MARTZ, R. C., STOOKEY, G. K., & FORNEY, R. B. (1976) Toxicity of sodium pentafluorostannite. A new anticariogenic agent. III. 30-day toxicity study in rats. *Toxicol. appl. Pharmacol.,* **35**: 21—28.

CORBIN, H. B. (1970) Separation and determination of trace amounts of tin present as organotin residues on fruits. *J. Assoc. Off. Anal. Chem.,* **53**: 140—146.

CORBIN, H. B. (1973) Rapid and selective pyrocatechol violet method for tin. *Anal. Chem.,* **45**: 534—537.

COSSA, P., DUPLAY, FISCHGOLD, ARFEL-CAPDEVIELLE, LAFON, PASSOUANT, MINVIELLE, & RADERMECKER, J. (1958) Encéphalopathies toxiques au stalinon. Aspects anatomocliniques et electroencéphalographiques. *Rev. Neurol.,* **98**: 97—108.

COTTON, F. A. & WILKINSON, G. (1972) *Advanced inorganic chemistry. A comprehensive text,* London, Wiley-Interscience, 3rd ed.

COTTON, F. A. & WILKINSON, G. (1976) *Basic inorganic chemistry,* New York, Wiley.

COUSSEMENT, S. (1972) Analyse des dépôts de fentin acétate sur feuilles de pomme de terre. *Ann. Gembloux,* **78**: 41—46.

COYLE, C. F. & WHITE, C. E. (1957) Fluorimetric determination of tin with flavonol. *Anal. Chem,* **29**: 1486—1488.

CREMER, J. E. (1957) The metabolism *in vitro* of tissue slices from rats given triethyltin compounds. *Biochem. J.,* **67**: 87—96.

CREMER, J. E. (1958) The biochemistry of organotin compounds: The conversion of tetraethyltin into triethyltin in mammals. *Biochem. J.,* **68**: 685—692.

CUTTER, H. C., FALLER, W. W., STOCKLEN, J. B., & WILSON, W. L. (1949) Benign pneumoconiosis in a tin oxide recovery plant. *J. ind. Hyg.,* **31**: 139—141.

DACK, G. M. (1955) Chemical poisons in food. In: *Food poisoning,* Chicago, University of Chicago Press, pp. 24—25.

DEPARTMENT OF NATIONAL HEALTH AND WELFARE, CANADA (1975) *Food and drug act and regulations office consolidation of the food and drugs act and the food and drugs regulations with amendments to December 18, 1975, B. 23.003,* Ottawa, Queen's Printer and Controller of Stationery, pp. 73B.

DEVJATYH, G. G., UMILIN, V. A., & CSINOVOI, Y. N. (1968) [Gas chromatographic analysis of tetrabutyltin.] *Tr. Himii i Himičeskoj Tehnol.,* No. d, pp. 82—85 (in Russian).

DITTRICH, T. & COTHERN, C. R. (1971) Analysis of trace metal particulates in atmospheric samples using X-ray fluorescence. *J. Air Pollut. Control Assoc.,* **21:** 716—719.

DRESSLER, M., MARTINU, V., & JANÁK, J. (1971) Response to the alkali flame ionization detector to silicon, tin and lead compounds. *J. Chromatogr.* **59:** 429—433.

DRESSLER, M., BARTH, M., & VESPALEC, R. (1975) Chromatographic determination of organotin compounds in fungistatic plaster. *Silikaty,* **4:** 301—310.

DRUAULT-TOUFESCO, M. N. (1955) A propos de deux cas d'intoxication grave par le Stalinon. *Bull. Soc. Ophtalmol. Fr.,* pp. 54—58.

DUCKERING, G. E. (1968) The cause of lead poisoning in the tinning of metals. *J. Hyg. (Lond.),* **8:** 474—503.

DUNDON, C. C. & HUGHES, J. P. (1950) Stannic oxide pneumoconiosis. *Am. J. Roentgenol.,* **63:** 797—811.

DURBIN, P. W. (1960) Metabolic characteristics within a chemical family. *Health Phys.,* **2:** 225—238.

DURFOR, C. N. & BECKER, E. (1964) Selected data on public supplies of the 100 largest cities in the United States, 1962. *J. Am. Water Works Assoc.,* **56:** 237—246.

DURUM, W. H. & HAFFTY, J. (1961) *Occurrence of minor elements in water,* Washington, DC, pp. 1—11 (Geological Survey Circular 445).

DYER, B. & TAYLOR, G. (1931) Notes from the Report of Public Analysts. *Analyst,* **56:** 251—253.

ELSEA, J. R. & PAYNTER, O. E. (1958) Toxicological studies on bis (tri-*n*-butyltin) oxide. *Am. Med. Assoc. Arch. ind. Health,* **18:** 214—217.

ELTEN, R. (1929) [Study on tin foil used in cheese packeting; rindless cheese.] *Chem.-Ztg.,* **53:** 586 (in German).

ENGBERG, Å. (1973) A comparison of a spectrophotometric (Quercetin) method and an atomic absorption method for the determination of tin in food. *Analyst,* **98:** 137—145.

EVANS, C. A. & MORRISON, G. H. (1968) Trace element survey analysis of biological materials by spark source mass spectrometry. *Anal. Chem.,* **40:** 869—875.

EVERETT, G. L., WEST, T. S., & WILLIAMS, R. W. (1974) The determination of tin by carbon filament atomic absorption spectrometry. *Anal. Chim. Acta,* **70:** 291—298.

EYRICH, W. (1972) [Determination of tin in canned fruits and vegetables.] *Dtsch. Lebensm.-Rundsch,* **9:** 280—282 (in German).

FEY, R. (1969) [Heavy metal content in canned fruits and vegetables: I. Tin and iron in canned asparagus.] *Ind. Obst-Geinneseverwert.,* **54:** 27—33 (in German).

FILER, T. D. (1971) Fluorometric determination of submicrogram quantities of tin. *Anal. Chem.,* **43:** 1753—1757.

FISCHER, H. W. & ZIMMERMAN, R. (1969) Long retention of stannic oxide: Lack of tissue reaction in laboratory animals. *Arch. Pathol.* **88:** 259—264.

FISH, R. H., KIMMEL, E. C., & CASIDA, J. E. (1975) Bioorganotin chemistry. Biological oxidation of tributyltin derivatives. *J. Organometal Chem.* **93** (1): C1—C4.

FISH, R. H., KIMMEL, E. C., & CASIDA, J. E. (1976) Bioorganotin chemistry: Biological oxidation of organotin compounds. In: Zuckerman, J. J., ed. *Organotin compounds: New chemistry and applications*, Washington, DC, American Chemical Society, pp. 197—203 (Advances in Chemistry Series 157).

FONTAN, M. M., VERGER, PERY, LOISEAU, & MULON (1955) Quatre cas dont deux mortels d'intoxication par le "Stalinon" chez l'enfant. *J. Med. Bordeaux*, **132**: 399—405.

FAO/WHO (1971) *1970 Evaluations of some pesticide residues in food*. Rome, Food and Agriculture Organization of the United Nations/World Health Organization, pp. 327—366, 521—542.

FOOD AND DRUG ADMINISTRATION, USA (1971) *Code of Federal Regulations, Title 21 — Foods and Drugs*, Washington DC, US Government Printing Office, pp. 409—410 (§ 121, 2602).

FREELAND, G. N. & HOSKINSON, R. M. (1970) Non-aqueous atomic absorption spectrophotometric determination of organometallic biocides. *Analyst*, **95**: 579—582.

FREITAG, K-D. & BOCK, R. (1974a) [Analysis of mixtures of tri-, di-, and monophenyltin compounds and inorganic tin (IV).] *Z. anal. Chem.*, **5**: 337—346 (in German).

FREITAG, K.-D. & BOCK, R. (1974b) Degradation of triphenyltin chloride on sugar beet plants. *Pesticide Sci.*, **5**: 731—739.

FURCHNER, J. E. & DRAKE, G. A. (1976) Comparative metabolism of radionuclides in mammals — XI. Retention of ^{113}Sn in the mouse, rat, monkey, and dog. *Health Phys.*, **31**: 219—224.

GAINES, T. B. & KIMBROUGH, R. D. (1968) Toxicity of fentin hydroxide to rats. *Toxicol. appl. Pharmacol.*, **12**: 397—403.

GATEHOUSE, B. M. & WILLIS, J. B. (1961) Performance of a simple atomic absorption spectrometer. *Spectrochim. Acta*, **17**: 710—718.

GAUER, W. O., SEIBER, J. N., & CROSBY, D. G. (1974) Determination of organotin residues from Plictran in fruit crops by gas-liquid chromatography. *J. agric. food Chem.*, **22**: 252—254.

GAUNT, I. F., COLLEY, J., GRASSO, P., & CREASEY, M., (1968) Acute and short-term toxicity studies on di-*n*-butyltin dichloride in rats. *Food Cosmet. Toxicol.*, **6**: 599—608.

GAYRAL, L., LAZORTHES, G., & PLANQUES, J. (1958) Méningo-encéphalite hypertensive par intoxication au Stalinon, guérie. *Rev. neurol.*, **98**: 143—144.

GEISSLER, H. & KRIEGSMANN, H. (1964) [Analysis of butyltin compounds $(n\text{-}C_4H_9)_nSnX_{4-n}$.] *Z. Chem.*, **9**: 354—355 (in German).

GEISSLER, H. & KRIEGSMANN, H. (1965) [Liquids used for separation in some gas chromatographic investigations of some n-butyltin compounds.]*Z. Chem.*, **11**: 423—424 (in German).

GELDMACHER - VON MALLINCKRODT, M. & POOTH, M. (1969) [Simultaneous determination of twenty five metals and metalloids in biological materials.] *Arch. Toxicol.*, **25**: 5—18 (in German).

GEORGE, G. M., ALBRECHT, M. A., FRAHM, L. J., & McDONNELL, J. P. (1973) Atomic absorption spectrophotometric determination of dibutyltin dilaurate in finished feeds. *J. Assoc. Off. Anal. Chem.*, **56**: 1480—1482.

GETZENDANER, M. E. & CORBIN, H. (1972) Residues on apples and pears from use of Plictran miticide. *J. agric. food Chem.*, **20**: 881—885.

GEYER, R. & ROTERMUND, U. (1969) [Oscillopolarographic determination of organotin compounds.] *Acta chim. Hung.*, **59** (2): 201—210 (in German).

GHAFOURI, R. (1970) Étude des eaux minérales d'Ax-les-Thermes. *Terres Eaux*, **64**: 19—27.

GOLDSCHMIDT, V. M. (1958) Tin, In: Muir, A. ed. *Geochemistry*, Oxford, Oxford University Press, pp. 391—397.

GORDON, M. (1952) [Trace elements in peat.] *Torfnachrichten,* **3:** 12 (in German).

DE GROOT, A. P. (1973) Subacute toxicity of inorganic tin as influenced by dietary levels of iron and copper. *Food Cosmet. Toxicol.,* **11:** 955—962.

DE GROOT, A. P., FERON, V. J., & TIL, H. P. (1973) Short-term toxicity studies on some salts and oxides of tin in rats. *Food Cosmet. Toxicol.,* **11:** 19—30.

GRUENWEDEL, D. W. & PATNAIK, R. K. (1973) Model studies regarding the internal corrosion of tin-plated food cans. IV. Polarographic investigation of the interaction of tin (II)-ions with L-cysteine. *Chem. Mikrobiol. Technol. Lebensm.,* **2:** 97—101.

GRUNER, J-E. (1958) Lésions du névraxe secondaires à l'ingestion d'éthylétain (Stalinon). *Rev. Neurol.,* **98:** 109—116.

GUARDASCIONE, V. & DI BOSCO, M. M. (1967) [Contribution to the knowledge of occupational disease due to pesticides; Three cases of acute intoxication by triphenyltin acetate fungicides.] *Lav. Um.,* **19:** 307—313 (in Italian).

HADŽIMICEV, P. (1971) [Natural content of heavy metals (copper, zinc, tin) in foods from the vegetable belt around Sofia.] *Khranit. Prom.,* **20:** 18—20 (in Bulgarian).

HALL, C. A. & LUDWIG, P. D. (1972) Evaluation of the potential use for several organotin compounds against the sheep blowfly *(Lucilia* spp.). *Vet. Rec.,* **90:** 29—32.

HAMILTON, T. A. (1948) The metabolic propartics of plutonium and allied material *Univ. Calif. Rad. Lab. Med. Health Phys. Q.,* pp. 4—16 (UCRL-270).

HAMILTON, E. I., MINSKI, M. J., CLEARY, J. J., & HALSEY, V. S. (1972) Comments upon the chemical elements present in evaporated milk for consumption by babies. *Sci. total Environ.,* **1:** 205—210.

HAMILTON, E. I., MINSKI, M. J., & CLEARY, J. J. (1972/1973) The concentration and distribution of some stable elements in healthy human tissues from the United Kingdom. An environmental study. *Sci. total Environ.,* **1:** 341—374.

HANNA, W. J. & GRANT, C. L. (1962) Spectrochemical analysis of the foliage of certain trees and ornamentals for 23 elements. *Bull Torrey Bot. Club,* **89:** 293—302.

HASEGAWA, T. & SUGIMAE, A. (1971) [Emission spectrographic determination on trace metals in airborne particulates.] *Jpn. Anal.,* **20:** 840—845 (in Japanese).

HAYASHI, H. (1969) [Square wave polarographic determination of dissolved tin in canned juice.] *Jpn. Anal.,* **18:** 58—61 (in Japanese).

HEATH, D. F. (1963) Some problems in the determination of residues in plants and animals. In: *Proceedings of a symposium; Radiation and Radioisotopes Applied to Insects of Agricultural Importance, Athens, 22—26 April 1963,* Vienna, International Atomic Energy Agency, pp. 185—193.

HEATH, D. F., (1967) The retention or triphenyltin and dieldrin and its relevance to the toxic effects of multiple dosing. In: *Radioisotopic detection of Pesticide Residues, Proceedings of a Panel, Vienna 1965,* pp. 18—26 (Abstract in: *Chem. Abstr.,* **67:** 1828 W).

HEDGES. T. R. & ZAREN, H. A. (1969) Experimental papilledema; A study of cats and monkeys intoxicated with triethyltin acetate. *Neurology, (Minneapolis),* **19:** 359—365.

HEINDL, R. A. (1970) Tin. In: *Mineral facts and problems, 1970 edition,* pp. 759—751. Reprint from: US Dept of the Interior, Bureau of Mines, Bulletin 650.

HEROK, J. & GÖTTE, H. (1963) [Radiometric analysis of triphenyltin acetate in the metabolism of plants and animals.] *Int. J. appl. Radiat. Isot.,* **14:** 461—479 (in German).

HEROK, J. & GÖTTE, H. (1964) [Radiometric study on the metabolism of triphenyltin acetate in milk sheep.] *Zentralbl. Vet.-Med. A,* **11:** 20—28 (in German).

HERON, P. N. & SPROULE, J. S. G. (1958) The chemical nature of fungicidal agents employed in the pulp and paper industry. *Indian Pulp Pap.,* **12:** 510—517.

HESLOP, R. B. & JONES, K. (1976) *Inorganic chemistry. A guide to advanced study.* Amsterdam, Elsevier.

HILES, R. A. (1974) Absorption, distribution and excretion of inorganic tin in rats. *Toxicol. appl. Pharmacol.,* **27:** 366—379.

HLEBNIKOVA, M. J. (1957) [Hygienic characteristics of dust in the production of tin.] *Sb. naucnyh Rab. voprogam Gig. Truda i profpatol.,* No. 2, pp. 53—83 (in Russian).

HODGE, V. F., SEIDEL, S. L., & GOLDBERG, E. D. (1979) Determination of tin (IV) and organotin compounds in natural waters, coastal sediments and macro algae by atomic absorption spectrometry. *Anal. Chem.,* **51:** 1259.

HOFF, J. E. & WILCOX, G. E. (1070) Accumulation of nitrate in tomato fruti and its effect on detinning. *J. Am. Soc. Hortic. Sci.,* **95:** 92—94.

HORÁČEK, V. & DEMČIK, K. (1970) [Group poisoning in spraying of field cultures with Brestan-60 triphenyltin acetate.] *Prac. Lék.,* **22:** 61—66 (in Czech).

HORIO, T. & NAKASEKO, M. (1972) [Spectrophotometric determination of dissolved tin in canned foods using Salicylidenamion-2-thiophenol.] *Tokyo Inst. Food Technol. J.,* **17:** 212—214 (in Japanese).

HORIO, T., IWAMOTO, Y., & SHIGA, I. (1967) The nitrate corrosion of canned orange juice drinks in Japan. Communication presented at the 5th *International Congress on Canned Fools, Vienna 3—6, October, 1967,* 12 pp.

HUEY, C., BRINCKMAN, F. E., GRIM, S., & IVERSON, W. P. (1974) The role of tin in bacterial methylation of mercury. Paper presented of the International Conference on Transport of Persistent Chemicals in Aquatic Ecosystems, Ottawa, Canada, 1—3 May, 1974.

IGNATJEVA, M. A., KUZNETSOV, I. G., MIRSKOV, R. G., & VLASOV, V. M. (1968) [Toxicology of acetylene organotin compounds of the triethylstannyl range.] *Izv. sib. Otdel. Akad. Nauk SSSR,* **1** (5): 118—119 (in Russian).

INNES, J. R. M., ULLAND, B. M., VALERIO, M. G., PETRUCELLI, L., FISHBEIN, L., HART, E. R., PALLOTTA, A. J., BATES, R. R., FALK, H. L., GART, J. J., KLEIN, .M, MITCHELL, I., & PETERS, J. (1969) Bioassay of pesticides and industrial chemicals for tumorigenicity in mice: A preliminary note. *J. Natl Cancer Inst.,* **42:** 1101—1114.

INTERNATIONAL LABOUR OFFICE (1972) Tin alloys and compounds. In: *Encyclopedia of Occupational Health and Safety,* Geneva, ILO, Vol. II, pp. 1407—1409.

IWAMOTO, Y., MIYAZAKI, M., KUNISATO, S., MAEDA, Y., HORIO, T., & KOMURA, S. (1968) [Studies on abnormal detinning of tin can with nitrate in foods: (II) Relationship between dissolution on tin and pH of canned tomato juice.] *J. Jpn Soc. Food Nutr.,* **21:** 50—54 (in Japanese).

JANITZ, W. (1971) [Comments on the polarographic method of tin and lead determination in tinned meat products.] *Med. Weter.,* **27:** 43—45 (in Polish).

JOHANSEN, O & STEINNES, E. (1969) Determination of tin in geological material by neutron activation analysis. *Analyst,* **94:** 976—978.

JOHNSON, J. H., YOUNGER, R. L., WITZEL, D. A., & RADELEFF, R. D. (1975) Acute toxicity of tricyclohexyltin hydroxide to livestock. *Toxicol. appl. Pharmacol.,* **31:** 66—71.

JOVANOVIĆ, D., STOJADINOVIĆ, L., & SPASOJEVIĆ, B. (1967) [Contribution to the determination of tin in canned meat.] *Vojno-sanit. Pregl.*, **24:** 405—408 (in Serbo-Croat).

KAHN, H. L. & SCHALLIS, J. E. (1968) Improvement of detection limits for arsenic, selenium and other elements with an argon-hydrogen flame. *At. Absorpt. Newsl.*, **7:** 5—9.

KANISAWA, M. & SCHROEDER, H. A. (1967) Life term studies on the effects of arsenic, germanium, tin, and vanadium on spontaneous tumors in mice. *Cancer Res.*, **27:** 1192—1195.

KANISAWA, M. & SCHROEDER, H. A. (1969). Life term studies on the effect of trace elements on spontaneous tumours in mice and rats. *Cancer Res.*, **29:** 892—895.

KAPPAS, A. & MAINES, M. D. (1976) Tin: A potent inducer of heme oxygenase in kidney. *Science*, **192:** 60—62.

KAS'YANENKO, I. V. & KUL'SKAYA, O. A. (1969) [Microelement content of the blood of patients with cancer of stomach and chronic gastritis secretory insufficiency.] *Vopr. eksp. Onkol Respub. Mezhvedom, SB*, **4:** 101—107 (in Russian).

KEENAN, R. G. & BYERS, D. H. (1952) Rapid analytical method for air pollution surveys. *Arch. ind. Hyg.*, **6:** 226—230.

KEHOE, R. A., CHOLAK, J. & STORY, R. V. (1940) A spectrochemical study of the normal ranges of concentration of certain trace metals in biological materials. *J. Nutr.*, **19:** 579—592.

KERK, VAN DER, G. J. M. (1975) [The present use of organotin compounds.] *Chem. Zeitung*, **99:** 26—32 (in German).

KIMBROUGH, R. D. (1976) Toxicity and health effects of selected organotin compounds: A review. *Environ. Health Perspect.*, **14:** 51—56.

KIMBROUGH, R. D. & GAINES, T. B. (1971) Hexachlorophene effects on the rat brain. *Arch. environ. Health*, **23:** 114—118.

KIMMEL, E. C., FISH, R. H., & CASIDA, J. E. (1977) Bio-organotin chemistry: Metabolism of organotin compounds in microsomal monoxygenase systems and in mammals. *J. Agric. Food Chem.*, **25** (1): 1—8.

KIMURA, Y., UEDA, K., NISHIZIMA, M., MAKI, T., WADA, M., TAKAHASHI, S., YAMADA, T., ARAI, M., & UEMATSU, T. (1970) [Studies on extractable heavy metals in canned foods: I. Detection of tin and lead from various sorts of canned baby foods after their opening.] *Ann. Rep. Tokyo Metrop. Inst. Hyg.*, **22:** 95—100 (in Japanese).

KIRK, R. S. & POCKLINGTON, W. D. (1969) The determination of tin and canned foods with quercetin. *Analyst*, **94:** 71—74.

KLEIN, A., COP, R., CHATRNÀ, E., & BADALOVÁ, M. (1970) [The study of vitamin C stability and growing of tin content in tinned juices.] *Česk. Hyg.*, **15:** 49—55 (in Czech).

KLEINKOPF, M. D. (1955) *Trace element exploration of Maine lake water*, Academic dissertation, Columbia University, 153 pp. (University Microfilm Pub. 12, 447).

KLEINKOPF, M. D. (1960) Spectrographic determination of trace elements in lake waters of northern Maine. *Geol. Soc. Am. Bull.*, **71:** 1231—1242.

KLIMMER, O. R. (1963) [Triphenyltin acetate and the toxicology of organic tin compounds.] *Arzneimittel-Forsch.*, **13:** 432—436 (in German).

KLIMMER, O. R. (1964) [Toxicological studies on triphenyltin acetate (TPZA).] *Zentralbl. Vet.-Med.*, **R.A 11:** 29—39 (in German).

KLIMMER, O. R. (1969) [The application of organotin compounds reviewed by the experimental toxicologist.] *Arzneimittel-Forsch.*, **19:** 934—939 (in German).

KOBAYASHI, T. & YADA, M. (1968) [Spectrophotometric determination of tin in foodstuffs with stilbazo.] *J. Food Hyg. Soc. Jpn*, **9:** 465—468 (in Japanese).

KOCH, J. & FIGGE, K. (1971) [Migration of tin stabilizers from PVC bottles into beer.] *Z. Lebensmittel-Unters.* **147**: 8—10 (in German).

KOČKIN, D. A., ŠKORBATOVA, J. C., PEGUSOVA, L. D., & VORONKOV, N. A. (1969) [Polarographic study on organotin compounds with an acid content.] *Zl. Obshch. Khim.*, **39**: 1777—1781 (in Russian).

KONOVALOV, G. S. & KOLESNIKOVA, T. K. (1969) [Trace elements in atmospheric precipitation around the Otkaznoe reservoir.] *Gidrokhim Mater.*, **49**: 74—79 (in Russian).

KRUTCHKOFF, D. J., JORDAN, T. H., WEI, S. H. Y., & NORDQUIST, W. D. (1972) Surface characterization of the stannous fluoride enamel interaction. *Arch. oral Biol.*, **17**: 923—930.

KRYLOVA, N. & BALABUH, A. (1970) [Trace element level in meat.] *Myas. Ind. SSSR*, 41: 39 (in Russian).

KUMPULAINEN, J. & KOIVISTOINEN, P. (1977) Advances in tin compound analysis with special reference to organotin pesticide residues. *Res. Rev.*, **66**: 1—18.

KUNISATO, S. & MIYAZAKI, M. (1972) [Application of plant growth regulators on fruits and vegetables for processing: I. Nitrate content in tomato fruits.] *J. Hortic. Assoc. Jpn.*, **41**: 411—420 (in Japanese).

KUROSAKI, N. & FUSAYAMA, T. (1973) Penetration of elements from amalgam into dentin. *J. dent. Res.*, **52**: 309—317.

KUTZNER, J. & BROD, K. H. (1971) [Resorption and excretion of tin after oral administration of tin-113.] *Nucl. Med. (Stuttg.)*, **10**: 286—297 (in German).

LAAMANEN, A., LÖFGREN, A., & PARTANEN, T. J. (1971) Some trace metals in particulates in the air of Helsinki and Turku, 1964—1969. *Work Environ. Health.*, **8**: 63—69.

LANDA, K., FEJFUŠOVÁ, J., & NEDOMLELOVÁ, R. (1973) [The hazards of organic tin compounds used as fungicides in some industrial applications.] *Prac. Lék.*, **25**: 391—394 (in Czech).

LANIGEN, D. & WEINBERG, E. L. (1976) The use of estertin stabilizers in PVC. In: Zucherman, J. J., ed. Organotin compounds: New chemistry and applications. Washington, DC, American Chemical Society (Advances in Chemistry Series 157).

LECOQ, R. (1954) Contribution à l'étude des effects toxiques du tétraéthylétain. *C. R. Hebd. Acad. Sci. (Paris)*, **239**: 678—680.

LEE, E. E., GORANSON, S. S., ENRIONE, R. E., & MORGAN, G. B. (1972) National air surveillance cascade impactor network. II. Size distribution measurements of trace metal components. *Environ. Sci. Technol.*, **6**: 1025—1030.

LINDEMANN, G., PINDEBORG, J. J., & PAULSEN, H. (1959) Recovery of the rat kidney in fluorosis. *Arch. Pathol.*, **67**: 30—33.

LJASKOVSKAJA, Y., & KRASIL'NIKOVA, T. (1961) [Photometric determination of tin in canned food by reaction with quercetin.] *Myasn. Ind. SSSR* **4**: 44—45 (in Russian) (English abstract in *Anal. Abstr.* (1963) **10**: 1177).

LOUNAMAA, J. (1956) Trace elements in plants growing wild on different rocks in Finland: A semi-quantitative spectrographic survey. *Ann. Bot. Soc. "Vanamo"* **29**: 4.

LOWRY, R. R. & TINSLEY, I. J. (1972) Determination of tin salts in fats. *J. Am. Oil. Chem. Soc.*, **49**: 508—509.

LUFF, A. P. & METCALFE, G. H. (1890) Four cases of tin poisoning caused by tinned cherries. *Br. med. J.*, **1**: 833—834.

LUKE, C. L. (1956) Photometric determination of tin with phenylfluorone: Determination of tin in lead and 1% antimony-lead alloys. *Anal. Chem.*, **28**: 1276—1279.

LYLE, W. H. (1958) Lesions of the skin in process workers caused by contact with butyltin compounds. *Br. J. ind. Med.*, **15**: 193—196.

MACINTOSH, R. M. (1968) Tin and tin alloys. In: *Kirk-Othmer encyclopedia of chemical technology,* 2nd ed. New York, Interscience pp. 273—325.

MAGEE, P. N., STONER, H. B., & BARNES, J. M. (1957) The experimental production of oedema in the central nervous system of the rat by triethyltin compounds. *J. Pathol. Bacteriol.,* **73**: 107—124.

MAHADEVAIAH, M., GOWRAMMA, R. V., RADAKRISHNAIAH SETTY, G., SASTRY, M. V., SASTRY, L. V. L., & BHATNAGAR, H. C. (1969) Studies on variation in tin content in canned mango nectar during storage. *J. food Sci. Technol.,* **6**: 192—196.

MAMONTOVA, A. A. (1940) [Effects of tin on animal organism.] *Vopr. Pitan,* **2**: 13—25 (in Russian).

MARHOLD, J. V. (1972) [*A collection of results of toxicological evaluation of substances and products.*] Pro Vychovu Vedoucicn Pracovniku Chemickeho Prumyclu Praha (in Czech).

MARKIĆEVIĆ, A. & TURKO, V. (1967) [Lesions due to triphenyltin acetate "Brestan".] *Arh. Hig. Rada,* **18**: 355—358 (in Serbo-Croat).

MARKS, M. J., WINEK, C. L., & SHANOR, S. P. (1969) Toxicity of triphenyltin hydroxide. (Vancide KS). *Toxicol. appl. Pharmacol.,* **14**: 627.

MASON, B. (1966) *Principles of geochemistry,* 3rd ed., New York, Wiley.

MAZAEV, V. T. & KOROLEV, A. A. (1969) [Experimental standardization of the maximum permissible concentration of the dichlorbutyltin in surface water.] *Prom. Zagrjaznenija Vodoemov,* No 9, pp. 15—37 (in Russian).

MAZAEV, V. T. & ŠLEPNINA, T. G. (1973) [Experimental data on hygienic standardization of dibutyltin sulfide in water bodies.] *Gig. i. Sanit.,* **8**: 10—13 (in Russian).

MAZAEV, V. T., LOSEV, N. I., & VOINOV, V. A. (1968) [On some vegetative reactions and on electrical activity of the brain cortex after intoxication with organic derivatives of tin.] *Bull. exp. Biol. Med.,* **11**: 72—74 (in Russian).

MAZAEV, V. T., KOROLEV, A. A., & SKAČKOVA, I. N. (1971) Experimental investigation of the toxic effect of tetraethyltin and dichlorodibutyltin on the animal organism. *J. Hyg. Epidemiol. Microbiol. Immunol.,* **15**: (1): 115—120.

MAZAEV, V. T., GOLOVANOV, O. V., IGUMOV, A. S., & TSAY, V. N. (1976) [Problem of the transformation of organotin compounds in a water medium.] *Gig. i Sanit,* **3**: 17—20 (in Russian).

MEINEMA, H. A., BURGER-WIERSMA, T., VERSLUIS-DE-HAAN, G. & GEVERS, E. Ch. (1978) Determination of trace amounts of butyltin compounds in aqueous systems by gas chromatography/mass spectrometry. *Environ. Sci. Technol.,* **12** (3): 288—293.

MIDWEST RESEARCH INSTITUTE (1977) Assessment of the need for the character of, and the impact resulting from limitations on selected organotins. Phase I Assessment of the need for limitations on organotins. Washington, DC, US Environmental Protection Agency, EPA Contract No. 68—01-4313).

MIJATOVIĆ, M. (1972) [Chronic hepatitis due to exposure to Brestan.] *Jugosl. inostr. Dok. Zast. Radu,* **8**: 3—9 (in Serbo-Croat).

MIKI, E. & FUKUI, Y. (1971) [Square wave polarographic determination of tin in canned fruit.] *Sci. Rep. Agric. Fac. Kagawa Univ.,* **22**: 35—40 (in Japanese).

MITCHELL, R. C. (1948) *The spectrographic analysis of soils, plants, and related materials.* Bureau of Soil Science, pp. 1—173 (Technical Communication No. 44).

MONIER-WILLIAMS, G. W. (1949) *Trace elements in food,* London, Chapman & Hall, pp. 138—161.

MOSKALEV, Ju. I. (1964) [Studies on the distribution of ^{113}Sn. In: *Distribution, biological effects and rapid excretion of radioactive isotopes.*] Moscow, Medicina, pp. 78—82 (in Russian).

MULAY, I. L., RAY, R., KNOX, B. E., SUHR, N. H., & DELANEY, W. E. (1971) Trace metal analysis of cancerous and non-cancerous human tissues. *J. Natl. Cancer Inst.*, **47**: 1—13.

NAKAHARA, T., MUNEMORI, M., & MUSHA, S. (1972) Atomic absorption spectrometric determination of tin in premixed inert gas (entrained air) hydrogen flames. *Anal. Chim. Acta*, **62**: 267—278.

NAKAMURA, Y. & KAMIWADA, S. (1973) [Colorimetric determination of tin by decomposition and tin-phenylfluorene complex.] *J. food Hyg.*, **14**: 352—356 (in Japanese).

NATIONAL ACADEMY OF SCIENCES. WASHINGTON (1973) *Toxicants occurring naturally in foods*, 2nd ed., Washington, DC, NAS, pp. 63—64.

NATIONAL INSTITUTE FOR OCCUPATIONAL SAFETY AND HEALTH (1976) *Criteria for a recommended standard... occupational Exposure to organotin compounds*, Washington, DC, US Government Printing Office, 187 pp.

NATIONAL WATER QUALITY NETWORK (1960) *Annual compilation of data, 1 October 1959—30 September 1960*, Washington, DC, US Department of Health, Education and Welfare, Public Health Service, pp. 353—356.

NEHRING, P. (1972) [Tin in peach preserves.] *Ind. Obst. Gemüsseverwert.*, **57**: 489—492 (in German).

NELSON, A. A. (1954) Approximate relation of parts per million in diet to mg/kg/day. *Assoc. Food Drug Off. Q. Bull*, **18**: 66.

NEUBERT, G. (1964) [Analysis of organotin stabilizers.] *Z. anal. Chem.*, **203**: 265—272 (in German).

NEUBERT, G. & ANDREAS, H. (1976) [The analysis of organotin compounds.] *Z. anal. Chem.*, **280**: 31 (in German).

NEUBERT, G. & WIRTH, H. O. (1975) [The analysis of organotin stabilizers.] *Z. anal. Chem.*, **273**: 19—23 (in German).

NEUMANN, W. P. (1970) *The organic chemistry of tin*, London, Wiley.

NEWMAN, E. J. & JONES, P. D. (1966) Separation and determination of small amounts of tin. *Analyst*, **91**: 406—410.

NIKONOROW, M., MAZUR, H., & PIEKACZ, H. (1973) Effects of orally administered plasticizers in the rat. *Toxicol. appl. Pharmacol.*, **26**: 253—259.

NISHIJIMA, M., UEDA, K., MAKI, T., WADA, M., TAKAHASHI, S., KIMURA, Y., YAMADA, T., NAKAJIMA, K., & FUKUDA, T. (1971) [Studies on heavy metals dissolved in canned foods. (II) Detection of tin and lead from various canned fruits after their opening.] *Ann. Rep. Tokyo Metrop. Inst. Hyg.*, **23**: 189—196, (in Japanese).

OBRUSNIK, I. (1969), Determination of indium and tin by activation analysis using replacement substoichiometry. *Talanta*, **16**: 563—566.

ONISHI, H. & SANDELL, E. B. (1957) Meteoritic and terrestrial abundance of tin. *Geochim. Cosmochim. Acta*, **12**: 262—270.

OYANGUREN, H., HADDAD, R., & MAASS, H. (1958) Stannosis. Benign pneumoconiosis owing to inhalation of tin dust and fume. I. Environmental and experimental studies. *Ind. Med. Surg.*, **27**: 427—431.

PAL, B. K. & RYAN, D. E. (1969) Fluorescence and metallic valency states. Part III. Determination of tin. *Anal. Chim. Acta*, **48**: 227—231.

PATE, B. D. & HAYS, R. L. (1968) Histological studies of testes in rats treated with certain insect chemosterilants. *J. econ. Entomol.* **61**: 32—34.

PEARLMAN, R. S., HEFFERREN, J. J., & LYON, H. W. (1970) Determination of tin in enamel and dentin by atomic absorption spectroscopy. *J. dent. Res.*, **49**: 1437—1443.

PEDLEY, F. G. (1927) Chronic poisoning by tin and its salts. *J. ind. Hyg.*, **9**: 43—47.

PELIKÁN, Z. & ČERNÝ, E. (1968a) [The toxic effect of tri-n-butyltin compounds on white mice.] *Arch. Toxikol.*, **23**: 283—292 (in German).

PELIKÁN, Z., & ČERNÝ, E. (1968b) The effect of low doses of bis- (tri-n-butyltin) oxide on the skin of rats. *Berufsdermatosen,* **16**: 340—349.

PELIKÁN, Z., & ČERNÝ, E. (1969) Toxic effect of bis- (tri-n-butyltin) oxide (TBTO) on the skin of rats. *Berufsdermatosen,* **17**: 305—316.

PELIKÁN, Z. & ČERNÝ, E. (1970a) The toxic effects of some di- and mono-n-octyltin compounds on white mice. *Arch. Toxikol.*, **26**: 196—202.

PELIKÁN, Z. & ČERNÝ, E. (1970b) Toxic effects of some mono-n-butyltin compounds on white mice. *Arch. Toxikol.*, **27**: 79—84.

PENDERGRASS, E. P. & PRYDE, A. W. (1948) Benign pneumoconiosis due to tin oxide. *J. ind. Hyg.*, **30**: 119—123.

PEREL'MAN, A. I. (1972) [Geochemistry of elements in the zone of hypergenesis.] Moscow, Nedra (in Russian).

PERRY, H. M. & PERRY, E. F. (1959) Normal concentrations of some trace metals in human urine: Changes produced by ethylenediamine tetra-acetate. *J. clin. Invest.*, **38**: 1452—1463.

PESME, P. (1955) Complicationes oculaires observées chez quatre enfants intoxiqués par le "Stalinon". *Arch. Fr. Pediat.*, **12**: 327—328.

PIVER, W. T. (1973) Organotin compounds: Industrial applications and biological investigation. *Environ. Health. Perspect.*, **4**: 61—79.

PORTRETNYJ, V. P., MALYUTA, V. F., & CHUIKO, V. T. (1973) [Determination of tin in water, after concentration by co-precipitation, by anodic stripping voltametry.] *Zh. anal. Khim.*, **28**: 1337—1340 (in Russian).

PRICE, W. J. & ROOS, J. T. H. (1969) Analysis of fruit juice by atomic absorption spectrophotometry. I — The determination of iron and tin in canned juice. *J. Sci. food Agric.*, **20**: 437—439.

PRÜLL, G. & ROMPEL, K. (1976) [Neurological and cerebro-electrical disturbances in acute poisoning due to organotin compounds.] *Nervenarzt*, **41**: 516—520 (in German).

REMY, H. (1956) *Treatise on inorganic chemistry,* Amsterdam, Elsevier, Vol. I.

RIDLEY, W. P., DIZIKES, L. J., & WOOD, J. M. (1977) Biomethylation of toxic elements. *Science,* **197**: 329—332.

RITTENHOUSE, G., FULTON, R. B., III GRABOWSKI, R. J., & BERNARD, J. L. (1969) Minor elements in oil field waters. *Chem. Geol.*, **4**: 189—209.

ROBERTS, R. M. G. & EL KAISSI, F. (1968) Base-catalyzed clearage of allylic derivatives of group IV A elements. *J. Organometal. Chem.*, **12 (1)**: 79—88.

ROBERTSON, A. J. (1960) Pneumoconiosis due to tin oxide. In: King, E. J. & Fletcher, C. M. ed *Symposium on industrial pulmonary diseases,* Boston, Little-Brown, pp. 168—184.

ROBERTSON, A. J. & WHITAKER, P. H. (1957) Radiological changes in pneumoconiosis due to tin oxide. *J. fac. Radiol.*, **6**: 224—233.

ROBINSON, I. M. (1969) Effects of some organotin compounds on tissue amine levels in rats. *Food Cosmet. Toxicol.*, **7**: 47—52.

ROE, F. J. C., BOYLAND, E., & MILLICAN, K. (1965) Effects of oral administration of two tin compounds to rats over prolonged periods. *Food Cosmet. Toxicol.*, **3**: 277—280.

RONDEPIERRE, J., TRUHAUT, R., GUILLY, P., HIVERT, P. E., & BARANDE, I. (1958) Sur une intoxication aiguë par un dérivé organiqué de l'étain. *Rev. neurol.*, **98**: 135—140.

ROSE, M. S. (1969) Evidence for histidine in the triethyltin-binding site of rat haemoglobin. *Biochem. J.*, **111**: 129—137.

ROSS, W. J. & WHITE, J. C. (1961) Application of pyrocatechol violet as a colorimetric reagent for tin. *Anal. Chem.*, **33**: 421—424.

SAITO, I. & ENDO, M. (1970) [Analytical studies on inorganic elements of gall stones.] *Hokkaido Igaku Zasshi,* **14**: 334—337 (in Japanese).

SATO, N., TSURUTA, K., KAMADA, I., NARITA, S., & ABUKAWA, H. (1973). [Studies on dissolved tin in canned foods by atomic absorption spectrophotometry.] *J. Food Hyg Soc. Ipn,* **14**: 245—248 in Japanese).

SAVAGE, W. (1939) Canned foods in relation to health, *Lancet,* **231** (2): 991—995.

SAWYER, R. (1967) Determination of dialkyltin stabilizers in aqueous extracts from PVC and other plastics. *Analyst,* **92**: 569—574.

SCHALLIS, J. E. & KAHN, H. L. (1968) The determination of tin in lubricating oils with a nitrous oxide-acetylene flame. *At. Absorpt. Newsl.,* **7**: 84.

SCHRAMEL, P., SAMSAHL, K., & PAVLU, J. (1973) Some determinations of Hg, As, Se, Sb, Sn, and Br in water plants, sediments and fishes in Bavarian rivers. *Int. J. environ. Stud.,* **5**: 37—40.

SCHROEDER, H. A. & BALASSA, J. J. (1961) Arsenic, germanium, tin and vanadium in mice: Effects on growth, survival and tissue levels. *J. Nutr.,* **92**: 245—253.

SCHROEDER, H. A., BALASSA, J. J., & TIPTON, I. H. (1964) Abnormal trace metals in man: *Tin. J. chron. Dis.,* **17**: 483—502.

SCHROEDER, H. A., KANISAWA, M., FROST, D. V., & MITCHENER, M. (1968) Germanium, tin and arsenic in rats: Effects on growth, survival, pathological lesions and life span. *J. Nutr.,* **96**: 37—45.

SCHROLL, E. VON & KRACHSBERGER, H. (1970) [Study on the geochemistry of impurities in atmospheric precipitates in the Vienna metropolitan area.] *Radex Rundsch.,* **5**: 331—341 (in German).

SCHRYVER, S. B. (1901) Some investigations on the toxicology of tin, with special reference to the metallic contamination of canned foods. *J. Hyg.,* **9**: 253—263.

SCHULER, P., CRUZ, E., GUIJON, C., MATURANA, V., & VALENZUELA, A. (1958) Stannosis. Benign pneumoconiosis owing to inhalation of tin dust and fume. II. Clinical study. *Ind. Med. Surg.,* **27**: 432—435.

SCHWARZ, K. (1971) *Tin as an essential growth factor for rats.* In: Mertz, W. & Comatzer, W. E., ed. *Newer trace elements in nutrition,* New York, Marcel Dekker, pp. 313—325.

SCHWARZ, K. (1974) Recent dietary trace element research, exemplified by tin, fluorine and silicon, *Fed. Proc.,* **33**: 1748—1757.

SCHWARZ, K., MILNE, D. B., & VINYARD, E. (1970) Growth effects of tin compounds in rats maintained in a trace element-controlled environment. *Biochem. biophys. Res. Commun.,* **40**: 22—29.

SCOTT, W. J. & STEWART, D. F. (1944) The influence of dissolved tin on the growth of *Clostridium botulinum* in canned vegetables. *J. Counc. Sci. Ind. Res. Aust.,* **17**: 16—22.

SEINEN, W. & WILLEMS, M. I. (1976) Toxicity of organotin compounds. I. Atrophy of thymus and thymus-dependent lymphoid tissue in rats fed di-n-octyltin dichloride. *Toxicol. appl. Pharmacol.,* **35**: 63—75.

SEINEN, W., VOS, J. G., SPANJE, I. van, SNOEK, M., BRANDS, R., & HOOYKAAS, H. (1977a) Toxicity of organotin compounds. II. Comparative *in vivo* and *in vitro* studies with various organotin and organolead compounds in different animal species with special emphasis on lymphocyte cytotoxicity. *Toxicol. appl. Pharmacol.,* **42**: 197—212.

SEINEN, W., VOS, J. G., KRIEKEN, R. van, PENNINKS, A., BRANDS, R., & HOOYKAAS, H. (1977b) Toxicity of organotin compounds, III. Suppression of thymus-dependent immunity in rats. by di-n-butyltin dichloride and di-n-octyltin dichloride. *Toxicol. appl. Pharmacol.,* **42**: 213—224.

SHELDON, A. W. (1975) Effects of organotin anti-fouling coatings on man and his environment. *J. Paint Technol.,* **47**: 54—58.

SHIRAISHI, Y., KUZUHARA, Y., & SUENAGA, S. (1972) [Determination of tin in food by atomic absorption spectrometry.] *J. Food Hyg. Soc. Jpn* **13**: 97—100 (in Japanese).

SHIRASU, Y. (1970) *Acute and sub-chronic toxicology studies of Dowco 213 in mice and rats.* Test report on the sub-acute toxicity of an insecticide, Dowco 213, Tokyo, Animal Pharmacology Institute of Physical and Chemical Research, 11 pp. (in Japanese).

SHONO, T. & MATSUMURA, Y. (1970) Organic synthesis by electrolysis VI. Anodic oxidation of arylcyclopropanes. *J. org. Chem.,* **35:** 4157—4160.

SKAČKOVA, I. N. (1967) [Hygienic substantiation of the maximum permissible concentration of tetraethyltin in water bodies.] *Gig. i Sanit.,* **4:** 11—17 (in Russian).

SKEEL, RL ,. & BRICKER, C. E. (1961) Spectrophotometric determination of dibutyltin dichloride. *Anal. Chem.,* **33:** 428—431.

SMITH, J. D. (1970) The spectrophotometric determination of microgram amounts of tin with phenylfluorone. *Analyst,* **95:** 347—350.

SODERQUIST, C. J. & CROSBY, D. G. (1978) Determination of triphenyltin hydroxide and its degradation products in water. *Anal. Chem.,* **50:** 1435—1439.

STANCULESCU, C., MARIN, E., SANDULESCU, C., ILLE, C., TRICIA, A., ILLIESCU, M., PETROVICI, C., & POPESCU, P. (1972) [Copper, lead, tin and arsenic content in canned fish after storage. II. Canned fresh water fish.] *Lucr. Inst. Cercet. Projiect. Aliment.,* **10:** 295—309 (in Romanian).

STOCKDALE, M., DAWSON, A. P., & SELWYN, M. J. (1970) Effects of trialkyltin and triphenyltin compounds on mitochondrial respiration. *Eur. J. Biochem.,* **15:** 342—351.

STONE, O. J., & WILLIS, C. J. (1968) The effect of stannous fluoride and stannous chloride on inflammation. *Toxicol. appl. Pharmacol.,* **13:** 332—338.

STONER, H. B. (1966) Toxicity of triphenyltin. *Br. J. ind. Med.,* **23:** 222—229.

STONER, H. B., BARNES, J. M., & DUFF, J. I. (1955) Studies on the toxicity of alkyl tin compounds. *Br. J. Pharmacol.,* **10:** 16—25.

STOROZEVA, H. H. (1963) [Contents of lead and tin in human teeth in normal and caries cases.] *Stomatologia,* **1:** 44—48 (in Russian).

STUDER, R. K., SIEGEL, B. A., MORGAN, J., & POTCHEN, E. J. (1973) Dexamethasone therapy of triethyltin induced cerebral edema. *Exp. Neurol.,* **38:** 429—437.

SUGIMAE, A. (1974) Emission spectrographic determination of trace elements in airborne particulate matter. *Anal. Chem.* **46:** 1123—1125.

SUZUKI, K. (1971) Some new observations in triethyltin intoxication of rats. *Exp. Neurol.,* **31:** 207—213.

SVENSSON, V. (1975) [Tin poisoning caused by canned peaches.] *Hyg. och Miljö,* No. 6, 25—27. (in Swedish).

TABOR, E. C. & WARREN, W. V. (1958) Distribution of certain metals in the atmosphere of some American cities. *Arch. ind. Hyg.,* **17:** 145—151.

THEUER, R. C., MAHONEY, A. W., & SARETT, H. P. (1971) Placental transfer of fluoride and tin in rats given various fluoride and tin salts. *J. Nutr.,* **101:** 525—532.

THOMPSON, K. C. & THOMERSON, D. R. (1974) Atomic absorption studies on the determination of antimony, arsenic, bismuth, germanium, lead, selenium, tellurium and tin by utilizing the generation of covalent hydrides. *Analyst,* **99:** 595—601.

TIHONOVA, Z. I. & ZORE, V. A. (1968) [Spectroscopic determination of manganese, copper, aluminium, lead and tin in certain vegetables and berries.] *Gig. i Sanit.,* **33:** 384—386 (in Russian).

TIPTON, I. H. (1960) The distribution of trace metals in the human body. In: Seven, M. T., ed. *Metal binding in medicine,* Philadelphia, Montreal, Lippincott, pp. 27—42.

TIPTON, I. H., COOK, M. J. STEINER, R. L., BOYLE, C. A., PERRY, H. M., & SCHROEDER, H. A. (1963) Trace elements in human tissue. Part I Methods. *Health Phys.,* **9:** 89—101.

TIPTON, I. H., STEWART, P. L., & MARTIN, P. G. (1966) Trace elements in diets and excreta. *Health Phys.,* **12:** 1683—1689.

TIPTON, I. H., STEWART, P. L., & DICKSON, J. (1969) Patterns of elemental excretion in long-term balance studies. *Health Phys.,* **16:** 455—462.

TJURIN, Y. M. & FLEROV, V. N. (1970) [Study of adsorption effects under electro-reducing of organotin compounds on mercury.] *Elektrochim.,* **6:** 1548—1552, (in Russian).

TJURIN, Y. M., FLEROV, V. N., & NIKITINA, Z. A. (1969) [On the mechanism of electro-reducing of the dialkyltin chloride on the mercury drip electrode.] *Elektrochim.,* **5:** 903 (in Russian).

TONGE, B. C. (1965) The gas chromatographic analysis of butyl- octyl-, and phenyltin halides. *J. Chromatogr.,* **19:** 182—184.

TORACK, R., GORDON, J., & PROKOP, J. (1969) Pathobiology of acute triethyltin intoxication. *Int. Rev. Neurobiol.,* **12:** 45—86.

TROMBETTI, G. & MAINI, P. (1970) Determination of tricyclohexyltin hydroxide residues in apples and pears. *Pestic. Sci.,* **1:** 144—149.

TRUHAUT, R., CHAUVEL, Y., ANGER, J. P., NGUYEN PHU LICH, VAN DEN DRIESSCHE, J., GUESNIER, L. R., & MORIN, N. (1976) Contribution to the toxicological and pharmacological study of tributyltin oxide. *Eur. J. Toxicol. environ. Hyg.,* **9** (1): 31—40.

TUTTLE, R. F., VOGT, J. R., & PARKINSON, T. F. (1970) Neutron activation analysis of trace elements in airborne particulates. In: *Proceedings of a Symposium on Nuclear Technology and Environmental Pollution, Salzburg,* Vienna, International Atomic Energy Agency, pp. 119—137 (Paper No. IAEA-SM-142a/4).

UDRIS, J. (1971) The analysis of tin stabilizers used in poly(vinylchloride) compositions. *Analyst,* **96:** 130—139.

UNITED NATIONS (1977) *United Nations yearbook of industrial satistics: Commodity production data* 1966—1975, Vol. II, pp. 498—500.

US ENVIRONMENTAL PROTECTION AGENCY (1973) *Air quality data for metals, 1968 and 1969 from the National Air Surveillance Networks,* Research Triangle Park, USEPA, June 1973.

VERNON, F. (1974) The fluorimetric determination of triphenyltin compounds. *Anal. Chim. Acta,* **71:** 192—195.

VINOGRADOV, A. P. (1953) *The elementary chemical composition of marine organisms,* Memoir Sears Foundation for Marine Research, New Haver, Yale University, pp. 16, 211, 240, 266, 360, 366, 535.

VINOGRADOV, A. P. (1956) Regularity of distribution of chemical elements in the earth's crust. *Geokhimiya,* **1:** 6—52.

VINOGRADOV, A. P. (1962) Average content of chemical elements in the principal types of igneous rocks of the earth's crust. *Geokhimiya,* 555—541.

VIZGIRDA, R. J. (1972) Organotin antifoulants. *Paint and varnish Produc.,* **62** (12): 12—28.

WALTERS, M. & ROE, F. J. C. (1965) A study on the effects of zinc and tin administered orally to mice over a prolonged period. *Food Cosmet. Toxicol.,* **3:** 271—276.

WARBURTON, S., UDLER, W., EWERT, R. M., & HAYNES, W. S. (1962) Outbreak of foodborne illness attributed to tin. *Public Health Rep. (Wash.),* **77:** 798—800.

WEAST, R. C., ed. (1974) *Handbook of Chemistry and Physics,* 55th ed. Cleveland, CRC Press.

WHO Technical Report Series, No. 515, 1973 *(Shistosomiasis control:* report of a WHO Expert Committee). pp. 30—32.

WHO (1975) *1974 evaluations of some pesticide residues in food,* pp. 155—158 (WHO Pesticides Residues Series No. 4).

WHO (1976) *Environmental Health Criteria 1: Mercury.* Geneva, WHO, 132 pp.

WIECZOREK, H. (1969) [The quantitative determination of organotin compounds by means of quercetin chelates.] *Dtsch. Lebensmitt—Rundsch.,* **3:** 74—78 (in German).

WILLIAMS, D. J. & PRICE, J. W. (1960) Paper chromatography of some organotin compounds. *Analyst,* **85:** 579—582.

WODSAK, W. (1967) Tin content of condensed milk. *Angew. Chem. Int. Ed.,* **6:** 885—886.

WOGGON, H. & JEHLE, D. (1973) [The inverse voltametric determination of biocidal organotin compounds.] *Die Nahrung,* **17:** 739—748 (in German).

WOGGON, H. & JEHLE, D. (1975) [The residual analysis of biocidal organotin compounds by means of inverse voltametric methods.] *Die Nahrung,* **19:** 271—275 (in German).

WOGGON, H., SÄUBERLICH, H., & UHDE, W. J. (1972) [A contribution to the determination (anodic stripping method) of organotin compounds and their degradation and decomposition products.] *Z. anal. Chem.,* **260:** 268—274 (in German).

WOIDICH, H. & PFANNHÄUSER, W. (1973) [Distribution of tin between the solid and liquid contents of cans of vegetables and fruits.] *Dtsch. Lebensmitt-Rundsch.,* **69:** 224—226 (in German).

WOOD, J. M. (1974) Biological cycles for toxic elements in the environment. *Science,* **183:** 1049—1052.

WOOD, J. M., SEGALL, H. J., RIDLEY, W. P., CHEH, A., CHUDYK, W., & THAYER, J. S. (1975) Metabolic cycles for toxic elements in the environment. In Hutchinson, T. C., ed. *Symposium Proceedings Vol. I, International Conference on Heavy Metals in the Environment. Toronto, Canada,* pp. 49—68.

WOOD, J. M., CHEH, A., DIZIKES, L. J., RIDLEY, W. P., RAKOW, S., & LAKOWICZ, J. R. (1978) Mechanisms for biomethylation of metals and metalloids. *Fed. Proc.,* **37:** 16—21.

YUM, N. M., CONINE, D. L., MARTZ, R. C., FORNEY, R. B., & STOOKEY, G. K. (1976) Renal tubular injury in rats induced by sodium pentalfluorostannite, a new anticariogenic agent. *Toxicol. appl. Pharmacol.,* **37:** 363—370.

ZOHM, H. (1972) [A new method for determination of tin and iron in canned vegetables and fruit juices.] Z. *Lebensmitt. -Unters.,* **149:** 30—33 (in German).

RA
565
A1
E58
VOL.
015

TIN AND ORGANOTIN
COMPOUNDS

8408

CIIT

P. O. Box 12137
Research Triangle Park, NC